GLENCOE
PHYSICS
Principles and Problems

Laboratory Manual

Teacher Edition

GLENCOE
McGraw-Hill

New York, New York   Columbus, Ohio   Woodland Hills, California   Peoria, Illinois

GLENCOE
PHYSICS
Principles and Problems

Student Edition

Teacher Wraparound Edition

Teacher Classroom Resources

Transparency Package with Transparency Masters

Laboratory Manual SE and TE

Physics Lab and Pocket Lab Worksheets

Study Guide SE and TE

Chapter Assessment

Tech Prep Applications

Critical Thinking

Reteaching

Enrichment

Physics Skills

Advanced Concepts in Physics

Supplemental Problems

Problems and Solutions Manual

Spanish Resources

Lesson Plans with block scheduling

Technology

Computer Test Bank (Win/Mac)

MindJogger Videoquizzes

Electronic Teacher Classroom Resources (ETCR)

Website at *www.glencoe.com/sec/science*

Physics for the Computer Age CD-ROM (Win/Mac)

The Glencoe Science Professional Development Series

Graphing Calculators in the Science Classroom

Cooperative Learning in the Science Classroom

Alternate Assessment in the Science Classroom

Performance Assessment in the Science Classroom

Lab and Safety Skills in the Science Classroom

Glencoe/McGraw-Hill

A Division of The McGraw-Hill Companies

Send all inquiries to:
Glencoe/McGraw-Hill
8787 Orion Place
Columbus, Ohio 43240

ISBN 0-02-825484-8
Printed in the United States of America.

7 8 9 045 05 04 03

Contents

denotes a Calculator Based Lab

DYO denotes a Design Your Own Lab

Physics: Principles and Problems

Contents *continued*

To the Teacher

The *Laboratory Manual* presents a flexible laboratory program that emphasizes the processes of science, the concepts of physics, and the attitudinal objectives of students. Students experience the principles of physics in action. The experiments are designed to promote enthusiasm for science via laboratory activities from which learners gain successful and observable results.

The skills of communicating experimental results to others through a laboratory report are developed as the student uses the report sheets that accompany each experiment. Most experiments are traditional and give definite results if the student uses reasonable care. Some experiments, called Design Your Own, give students the opportunity to formulate their own hypotheses and develop their own procedures to solve problems. These experiments, too, give definite results but allow for student creativity in attaining them. Some experiments are calculator-based. They take advantage of CBL technology so that students can easily measure, manipulate, and graph data. The variety of experiments provided in this *Laboratory Manual* helps ensure that your students will experience challenge and success in the physics laboratory.

Teacher notes are printed in the back of the Teacher Edition of the *Laboratory Manual* for easy reference. These include introductory information, science process skills, preparation hints, optional materials, and procedure hints. Complete answers and sample data are furnished for each experiment. Suggestions are given for evaluating student data and for determining the degree of precision to expect in student analysis of data. Sample graphs are provided for all data that students are required to plot.

The *Laboratory Manual* was written to accompany *Physics: Principles and Problems.* However, it is well suited for use with any current physics text. The 53 experiments allow for a wide range of choice in selecting those activities best suited to individual class or student needs.

The lab program is interactive with the Lab Partner™ graphing program available for IBM- or Apple II-compatible computers. Data can be easily plotted using the Lab Partner™ Program.

Materials List

Quantity needed for 12 lab stations or 24 students

Item	Quantity	Used in Experiment(s)
acrylic rod	12	20-1
alpha source	1	31-1
aluminum sheet, 5 cm × 5 cm	10	31-1
ammeter, 0–50 mA or 0–100 mA	12	23-1, 23-2, 27-1, 29-1
ammeter, 0–5 A	24	22-2, 24-2, 24-3, 26-1
ammeter (microammeter), 0–100 μA	12	27-2
apron	24	28-2
balance	1	6-1, 9-1, 12-1, 22-2, 24-3
battery, 1.5 V, C size	24	27-1, 29-1
battery, 9 V	12	29-2
battery clip, 9 V	12	29-2
battery holder, C size	24	27-1
beaker, 100 mL	12	1-1, 25-2
beaker, 150 mL	36	1-1, 2-2
beaker, 250 mL	24	12-1
beaker, 500 mL	12	13-1
beta source	1	30-1, 31-1
bicycle air pump	1	19-2
board, wooden	12	6-2, 7-1, 11-1
breadboard	12	29-2
burner, laboratory	12	28-2
calculator, TI-82 with unit-to-unit link cable	12	6-3, 12-2
calculator, TI-92	12	6-1, 6-3, 7-1
capacitor, 1000 μF	12	21-1
carbon-paper disks	12	5-2
carbonated water, 100 mL	12	1-1
cardboard, $8\frac{1}{2}$ × 11 inches	12	24-2
cardboard sheet, 5 cm × 5 cm	10	31-1
cardboard, with slit	12	28-2
cart, collision, one with a spring-release mechanism	24	9-1
cart, laboratory	12	5-1, 6-1
cassette tape of music	12	15-1
CBL unit	12	2-2, 3-1, 5-1, 6-1, 6-2, 6-3, 9-1, 12-1, 12-2, 16-2, 21-1, 22-1, 23-1, 26-1, 29-1
CBL DIN adapter	24	6-2, 22-1, 23-1, 29-1
centimeter ruler	12	5-2
clamp, C	24	3-1, 5-1, 5-2, 6-1
clamp, meterstick	36	7-3
clamp, ring-stand	24	7-2, 16-2, 24-2, 25-2
clay, modeling	12	28-2
clock	1	1-1
coffee can, large	12	12-2
coffee can, small	12	12-2
compass, small magnetic	12	24-1, 24-2
constant-velocity vehicle	12	3-1
corkboard	12	17-1
correction fluid, white, bottle	12	15-1
cross support, with rings or hooks	12	4-1
cup, paper	12	7-2, 24-3
cup, polystyrene	12	12-1, 13-1, 22-2
diffraction grating	12	28-2
diode, 1N4004	12	29-1
dowel rod, 30 cm long × 2 cm diam	12	14-1
dowel rod, various diameters	36	26-1
electronic timer with photogates	12	7-2
electroscope, vane or leaf	12	20-1
eyedropper	12	14-1
film, double slit	12	19-1
filters, red, orange, green, blue, yellow, violet	12	16-1, 19-1, 27-2
force sensor	12	6-2
force table	12	4-1
galvanometer, with centered zero	12	25-1, 25-2
gamma source	1	30-1, 31-1
Geiger counter or scaler	1	30-1, 31-1
glass plate, flat	12	14-1, 16-1, 17-2
glass rod	12	12-1, 20-1
glass, small pane	12	12-2
gloves, heat-protective	12	2-2, 16-2, 30-1
gloves, rubber	12	19-2
goggles, safety	24	2-2, 9-1, 9-2, 12-1, 16-1, 28-2

Item	Quantity	Used in Experiment(s)
graduated cylinder, 10 mL	36	2-2
graph paper	48	5-1, 15-1, 30-1
graph paper, polar	24	8-1
graphing calculator with PHYSICS CBL software program	12	2-2, 3-1, 5-1, 6-2, 9-1, 12-1, 21-1, 22-1, 23-1, 26-1, 29-1
gyroscopic bicycle wheel	12	9-2
helping-hand project holder	1	19-2
holders, for light, mirror, screen	12	18-1, 18-2
holographic film	12	19-2
holographic film developer kit	1	19-2
hot plate, electric	12	2-2, 12-1
Iceland spar calcite crystal	12	16-1
index card, 5 in. × 7 in.	24	9-1, 19-1, 19-2
inner tube, automobile	1	19-2
integrated circuits (CMOS): 4081, 4071, 4001, 4011	12	29-2
iron filings, bottle	1	24-1, 24-2
iron BBs, jar	12	24-3
iron nail	12	24-1, 24-2
knife switch	24	22-1, 23-1, 23-2, 24-2, 27-1, 29-1
knife switch, SPDT	12	21-1
lamp	12	16-2
laser, He-Ne	1	17-1, 18-2, 19-1, 19-2
lead sheet, 5 cm × 5 cm	10	31-1
lens, concave	12	18-2
lens, convex	12	18-2
Leydon jar	12	20-1
lightbulb, colored	12	16-2
lightbulb, 40 W, clear filament with socket and cord	12	16-1, 19-1, 27-2, 28-1
lightbulb, white, different wattages	48	16-2
light source, for lens/mirrors	12	18-1, 18-2
light source, for ripple tank	12	14-1, 14-2
light-emitting diode (LED)— red, green, yellow	12	27-1, 29-2
lighter	12	28-2
light probe	12	16-2
link cable	12	2-2, 3-1, 5-1, 6-2, 9-1, 12-1, 21-1, 22-1, 23-1, 26-1, 29-1

Item	Quantity	Used in Experiment(s)
magnet, bar	24	24-1, 25-1
magnet, disk	12	19-2
magnet, horseshoe	24	25-2
magnetic field sensor	12	23-1
magnifier, with scale	12	19-1
mass, 100 g, hooked	12	13-1
mass, 250 g	24	6-1
mass, 1000 g, hooked	12	6-1
mass, 500 g, hooked	12	4-1, 6-1, 10-2, 13-1, 16-1
mass, 1 kg	12	5-2
mass, 3 kg	24	9-2
mass-of-the-electron apparatus	12	26-1
mass set, hooked	12	10-1
mass set, small masses	12	6-1, 9-1
metal strip (parabola)	12	14-1
meterstick	24	2-1, 4-2, 6-1, 6-2, 6-3, 7-2, 7-3, 9-1, 10-1, 10-2, 14-1, 14-2, 15-2, 18-1, 18-2, 19-1, 24-2, 28-2, 30-1
meterstick support	24	18-1, 18-2
metric ruler, grooved	24	1-1, 3-1, 4-1, 6-1, 7-1, 7-3, 8-1, 11-1, 14-2, 17-1, 17-2, 18-1, 18-2, 19-1
mirror, concave	12	18-1, 19-2
mirror, convex	12	18-1
mirror, plane surface (flat and thick)	12	16-2
mirror, plane surface (flat and thin)	12	16-1, 17-1
newspaper, sheet	12	12-2
open-pipe tube, adjustable, 0.25–1.4 m long	12	15-2
oscilloscope	1	29-1
paper, black	12	19-2
paper, carbon	12	7-1
paper, lined	12	7-1
paper, plain white	48	4-1, 14-1, 14-2, 17-1, 17-2, 24-1, 24-2, 28-2
paper clips, steel, box	1	24-2
paper towel	12	13-1

Item	Quantity	Used in Experiment(s)
photoelectric-effect module, with amplifier	12	27-2
pins, straight	24	17-1
pith ball	12	20-1
plastic, clear object such as box or protractor	12	16-1
plastic tray, 10 cm × 10 cm	3	19-2
plastic wrap, sheet	12	12-2, 20-1
polarizing filter	24	16-1
polyethylene, sheet	12	12-2
potentiometer, 1000 Ω (variable resistor)	12	27-1, 29-1
power supply, 0–15 VDC	12	14-1, 14-2, 21-1, 22-2, 23-1, 23-2
power supply, 0–5 A	24	24-2, 24-3, 26-1
power supply, 0–5 VDC or two 1.5 V batteries	12	22-1
power supply, 90–250 VDC	12	26-1
power supply, 6 VAC or AC adapter, such as that used for a tape recorder	12	29-1
power supply, DC or AC filament, 6V	12	26-1
protractor	12	4-1, 6-2, 6-3, 14-1, 17-1, 17-2
pulley, double-mounted	24	10-1
pulley, single-mounted	24	5-1, 6-1, 10-1
pulley support	12	10-1
radio, portable, with earphones	12	15-1
raisins	48	1-1
recording timer with power supply	12	5-2
resistor, 10 Ω	12	29-1
resistor, 22 Ω	12	27-1
resistor, 1000 Ω	12	29-1, 29-2
resistor, 10 kΩ	12	21-1
resistor, 27 kΩ	12	21-1
resistors, 100 Ω, 150 Ω, 220 Ω, 470 Ω, 0.5-W rating	12	22-1, 23-1
resistors, 180 Ω, 220 Ω, 330 Ω, 0.5-W rating	12	23-2
rheostat	12	22-2
ring stand	24	4-1, 7-2, 9-1, 16-2, 24-2, 25-2
ripple tank	12	14-1, 14-2

Item	Quantity	Used in Experiment(s)
rubber bands, assorted sizes	48	17-1, 19-1
rubber rod	12	20-1
safelight, green	1	19-2
sandpaper, fine 2 cm × 2 cm	12	25-2
screen, paper or cardboard	12	18-1, 18-2
silicon spray, can	1	11-1
silk pad	12	20-1
sodium carbonate salts, anhydrous, sample	12	28-2
software, L06LIGHT program for calculator	12	16-2
solenoid, air-core	12	26-1
sound level meter	1	15-1
specific heat set	12	12-1
spectroscope	12	28-1
spectrum tubes: argon, bromine, carbon dioxide, chlorine, helium, hydrogen, krypton, mercury vapor, neon, nitrogen, oxygen, xenon	1	28-1
spectrum-tube power supply	1	28-1
spoon, plastic	12	1-1
spring scale, 5-N capacity	36	4-1, 5-2, 7-3, 10-1, 11-1, 13-1
stand, for Geiger-Mueller tube	1	31-1
steel ball	24	7-1, 7-2
steel plate, 30 cm × 30 cm	1	19-2
stool, rotating	12	9-2
stopwatch or watch with second hand	12	2-2, 7-1, 7-2, 14-2, 19-2, 22-2, 30-1, 31-1
string, heavy, various lengths	12	4-1, 5-1, 5-2, 6-1, 6-2, 6-3, 7-2, 10-1, 11-1, 12-1
stroboscope, hand	12	14-2
tape, masking	12	5-1, 5-2, 6-1, 6-2, 6-3, 7-1, 7-2, 7-3, 9-1, 14-2, 15-1, 18-1, 19-1, 24-2, 24-3, 25-2, 28-2, 30-1

Item	Quantity	Used in Experiment(s)	Item	Quantity	Used in Experiment(s)
temperature probe	12	2-2, 12-1	voltmeter, 0–250 VDC	12	26-1
thermal mitt	1	28-1	washer	144	7-2, 14-1
thermometer, $-210°$–$110°C$	48	2-2, 15-2, 22-2	wave generator, 2-point source	1	14-1, 14-2
thumbtacks	48	17-1			
TI-Graph Link	12	12-2	wax paper, sheet	12	12-2
TI temperature probe	12	12-2	wire, coil, 1-, 25-,100-turn	12	25-1
timer tape, roll	12	5-2	wire, connecting, 8 pieces	12	21-1, 22-1, 22-2, 23-1, 23-2, 24-3, 25-1, 25-2, 26-1, 27-1, 29-1, 29-2
tongs, beaker	12	2-2			
tongs, plastic	12	19-2			
tube, cardboard	12	25-2			
tuning-fork hammer	12	15-2			
tuning forks, various frequencies	24	15-2	wire, 14–18 gauge, 1.5 m	12	24-2
tweezers	12	30-1, 31-1	wire, 22–28-gauge enameled magnet, 60 m	12	25-2
U-shaped channel, 50 cm or longer	24	7-2	wire, insulated magnet, 2 m	12	24-3
			wire, with alligator clips	12	20-1
ultrasonic motion detector	12	3-1, 9-1	wood block	12	2-1, 7-1, 17-1
vernier motion detector	1	6-1	wood block, with hook in one end	12	11-1
vernier student force sensor with CBL DIN adapter	1	6-3			
voltage and current probe with dual-channel amplifier box	12	22-1, 23-1, 29-1	wood (or paraffin) block	36	14-1
			wool pad	12	20-1
voltage probe	12	21-1			
voltmeter, 0–5 VDC or 0–15 VDC	12	22-2, 23-1, 23-2, 27-1, 27-2, 29-1			

Note on CBL products

Sources of CBL units and probes used for this Lab Manual are:

Vernier Software
8565 SW Beaverton-Hillsdale Highway
Portland, OR 97225-2429
(503)297-5317

PASCO
P.O. Box 619011
10101 Foothills Blvd.
Roseville, CA 95678-9011
(916)786-3800
U.S. (800)772-8700

AccuLab Products
614 Scenic Drive
Suite # 104
Modesto, CA 95350
(209)522-8874

The PHYSICS program comes from Texas Instruments, T.I., and is often available with the CBL unit. It comes with eight subprograms (PHZCALIB, PHZCALS, PHZGRAPH, PHZMICRO, PHZMONIT, PHZOPTIO, PHZTIMEG, and PHZTRIGG). All of the programs should be loaded into the calculator. When PHYSICS is run, it will call the appropriate program for the sensor being used. The PHYSICS program uses a series of menus to guide the user through setup and calibration of probes. The program TIMER should also be loaded. These programs can be downloaded from the TI web site at *www.ti.com*. The program group PHYSICST contains the PHYSICS program and all of the subprograms. The TI-Graph Link is necessary to send files from a PC or Mac to the graphing calculator. There are different program groups depending upon the graphing calculator being used. Programs for the TI-82, 83, 85, 86, and 92 calculators are available from the web site. Additionally, programs are available from the Vernier Software web site at *www.vernier.com*.

If the CBL unit and graphing calculator do not work at first, unplug and replug the cables, pressing them into the units firmly.

GLENCOE

PHYSICS
Principles and Problems

● ● ● ● ● ● ● ● ● ● ● ● ●

Laboratory Manual
Student Edition

McGraw-Hill

New York, New York   Columbus, Ohio   Woodland Hills, California   Peoria, Illinois

GLENCOE

PHYSICS

Principles and Problems

Student Edition

Teacher Wraparound Edition

Teacher Classroom Resources

Transparency Package with Transparency
  Masters
Laboratory Manual SE and TE
Physics Lab and Pocket Lab Worksheets
Study Guide SE and TE
Chapter Assessment
Tech Prep Applications
Critical Thinking
Reteaching
Enrichment
Physics Skills
Advanced Concepts in Physics
Supplemental Problems
Problems and Solutions Manual
Spanish Resources
Lesson Plans with block scheduling

Technology

Computer Test Bank (Win/Mac)
MindJogger Videoquizzes
Electronic Teacher Classroom Resources
  (ETCR)
Website at *www.glencoe.com/sec/science*
Physics for the Computer Age CD-ROM
  (Win/Mac)

The Glencoe Science Professional Development Series

Graphing Calculators in the Science Classroom
Cooperative Learning in the Science Classroom
Alternate Assessment in the Science Classroom
Performance Assessment in the Science
  Classroom
Lab and Safety Skills in the Science Classroom

Glencoe/McGraw-Hill

A Division of *The McGraw·Hill Companies*

Copyright © by the McGraw-Hill Companies, Inc. All rights reserved. Permission is
granted to reproduce the material contained herein on the condition that such
material be reproduced only for classroom use; be provided to students, teachers,
and families without charge; and be used solely in conjunction with the *Physics:
Principles and Problems* program. Any other reproduction, for use or sale, is
prohibited without prior written permission of the publisher.

Send all inquiries to:
Glencoe/McGraw-Hill
936 Eastwind Drive
Westerville, Ohio 43081

ISBN 0-02-825483-X
Printed in the United States of America.

1 2 3 4 5 6 7 8 9 045 05 04 03 02 01 00 99 98

Contents

denotes a Calculator Based Lab
DYO denotes a Design Your Own Lab

Contents *continued*

Physics: Principles and Problems

To the Student

The *Laboratory Manual* contains 53 experiments for the beginning student of physics. The experiments illustrate the concepts found in this introductory course. Both qualitative and quantitative experiments are included, requiring manipulation of apparatus, observation, and collection of data. The experiments are designed to help you utilize the processes of science to interpret data and draw conclusions.

The laboratory report is an important part of the laboratory experience. It helps you learn to communicate observations and conclusions to others. Special laboratory report pages are included with each experiment to allow the most efficient use of lab-report time. Graph paper is necessary for most labs requiring construction of graphs.

While accuracy is always desirable, other goals are of equal importance in laboratory work that accompanies early courses in science. A high priority is given to how well laboratory experiments introduce, develop, or make the physics theories learned in the classroom realistic and understandable and to how well laboratory investigations illustrate the methods used by scientists. The investigations in the *Laboratory Manual* place more emphasis on the implications of laboratory work and its relationship to general physics principles, rather than to how closely results compare with accepted quantitative values.

Processes of Science

The scientifically literate person uses the processes of science in making decisions, solving problems, and expanding an understanding of nature. The *Laboratory Manual* utilizes many processes of science in all of the lab activities. Throughout this manual, you are asked to collect and record data, plot graphs, make and identify assumptions, perform experiments, and draw conclusions. In addition, the following processes of science are included in the *Laboratory Manual*.

Observing: Using the senses to obtain information about the physical world.

Classifying: Imposing order on a collection of items or events.

Communicating: Transferring information from one person to another.

Measuring: Using an instrument to find a value, such as length or mass, that quantifies an object or event.

Using Numbers: Using numbers to express ideas, observations, and relationships.

Controlling Variables: Identifying and managing various factors that may influence a situation or an event, so that the effect of any given factor may be learned.

Designing Experiments: Performing a series of data-gathering operations that provide a basis for testing a hypothesis or answering a specific question.

Defining Operationally: Producing definitions of an object, a concept, or an event in terms that give it a physical description.

Formulating Models: Devising a mechanism or structure that describes, acts, or performs as if it were a real object or event.

Inferring: Explaining an observation in terms of previous experience.

Interpreting Data: Finding a pattern or meaning inherent in a collection of data, which leads to a generalization.

Predicting: Making a projection of future observations based on previous information.

Questioning: Expressing uncertainty or doubt that is based on the perception of a discrepancy between what is known and what is observed.

Hypothesizing: Explaining a relatively large number of events by making a tentative generalization, which is subject to testing, either immediately or eventually, with one or more experiments.

The Experiment

Experiments are organized into several sections. Most of the experiments are traditional in nature. They open with a review or an introduction of relevant physics concepts and background information. Objectives listed in the margin help focus your inquiry.

The Materials section lists the equipment used in the experiment and allows you to assemble the required materials quickly and efficiently. All of the equipment listed is either common to the average high school physics lab or can be readily obtained from local sources or a science supply company. Slight variations of equipment can be made and will not affect the basic integrity of the experiments in the *Laboratory Manual*. Safety symbols alert you to potential dangers in the laboratory investigation.

The Procedure section contains step-by-step instructions to perform the experiment. This format helps you take advantage of limited laboratory time. Caution statements are provided where appropriate.

The Data and Observations section helps organize the lab report. All tables are outlined and properly labeled. In more qualitative experiments, questions are provided to guide your observations.

In the Analysis and Conclusions section, you relate observations and data to the general principles outlined in the objectives of the experiment. Graphs are drawn and interpreted and conclusions concerning data are made. Questions relate laboratory observations and conclusions to basic physics principles studied in the text and in classroom discussions.

The Extension and Application section allows you to extend and apply some aspect of the physics concept investigated. This section may include supplemental procedures or problems that expand the scope of the experiment. They are designed to further the investigation and to challenge the more interested student. Often this section illustrates a current application of the concept.

Some of the experiments are called Design Your Own. Their format is similar to the Design Your Own labs in your textbook. As with the traditional labs, they begin with introductory information and objectives. A statement of the problem focuses the challenge for the experiment. The Hypothesis section reminds you to use what you know to develop a possible explanation of the problem. You then have the opportunity to develop your own procedure to test your hypothesis. The Plan the Experiment section provides overall guidance for this process. The list of materials includes items that could be used for the experiment, depending on your procedure. You may choose to use all, some, or none of these items. Your teacher will help you determine the safe use of materials when he or she checks your procedure. In most cases, a table is provided in which you can record your data. Analyze and Conclude questions help you make sense of your data and determine whether or not they support your hypothesis. Finally, Apply questions give you the opportunity to apply what you've learned to new situations.

Introduction

Purpose of Laboratory Experiments

The laboratory work in physics is designed to help you to better understand basic principles of physics. You will, at the same time, gain a familiarity with the scientific methods and techniques employed in the laboratory. In each experiment, you will be seeking a definite goal, investigating a specific principle, or solving a definite problem. To find the answer to the problem, you will make measurements, list your measurements as data, and then interpret the data to find the results of your measurements.

The values you obtain may not always agree with accepted values. Frequently this result is to be expected because your laboratory equipment is usually not sophisticated enough for precision work, and the time allowed for each experiment is not extensive. The relationships between your observations and the broad general laws of physics are of much more importance than strict numerical accuracy.

Preparation of Your Lab Report

One very important aspect of laboratory work is the communication of your results obtained during the investigation. This laboratory manual is designed so that laboratory report writing is as efficient as possible. In most of the laboratory experiments in this book, you will write your report on the report sheets placed immediately following each experimental procedure. All tables are outlined and properly labeled for ease in recording data and calculations. Adequate space is provided for necessary calculations, discussion of results, conclusions, and interpretations. For instructions on writing a formal laboratory report, see page xix.

Using Significant Digits

When making observations and calculations, you should stay within the limitations imposed upon you by your equipment and measurements. Each time you make a measurement, you will read the scale to the smallest calibrated unit and then obtain one smaller unit by estimating. The doubtful or estimated figure is significant because it is better than no estimate at all and should be included in your written values.

It is easy, when making calculations using measured quantities, to indicate a precision greater than your measurements actually allow. To avoid this error, use the following guidelines:

- When adding or subtracting measured quantities, round off all values to the same number of significant decimal places as the quantity having the least number of decimal places.

- When multiplying or dividing measured quantities, retain in the product or quotient the same number of significant digits as in the least precise quantity.

Accuracy and Precision

Whenever you measure a physical quantity, there is some degree of uncertainty in the measurement. Error may come from a number of sources, including the type of measuring device, how the measurement is made, and how the measuring device is read. How close your measurement is to the accepted value refers to the accuracy of a measurement. In several of these laboratory activities, experimental results will be compared to accepted values.

When you make several measurements, the agreement, or closeness, of those measurements refers to the precision of the measurement. The closer the measurements are to each other, the more precise is the measurement. It is possible to have excellent precision but to have inaccurate results. Likewise, it is possible to have poor precision but to have accurate results when the average of the data is close to the accepted value. Ideally, the goal is to have good precision and good accuracy.

Relative Error

While absolute error is the absolute value of the difference between your experimental value and the accepted value, relative error is the percentage deviation from an accepted value. The relative error is calculated according to the following relationship:

$$\text{relative error} = \frac{|\text{accepted value} - \text{experimental value}|}{\text{accepted value}} \times 100\%.$$

Graphs

Frequently an experiment involves finding out how one quantity is related to another. The relationship is found by keeping constant all quantities except the two in question. One quantity is then varied, and the corresponding change in the other is measured. The quantity that is deliberately varied is called the independent variable. The quantity that changes due to the variation in the independent variable is called the dependent variable. Both quantities are then listed in a table. It. is customary to list the values of the independent variable in the first column of the table and the corresponding values of the dependent variable in the second column.

More often than not, the relationship between the dependent and independent variables cannot be ascertained simply by looking at the written data. But if one quantity is plotted against the other, the resulting graph gives evidence of what sort of relationship, if any, exists between the variables. When plotting a graph, use the following guidelines:

- Plot the independent variable on the horizontal x-axis (abscissa).

- Plot the dependent variable on the vertical y-axis (ordinate).

- Draw the smooth line that best fits the most plotted points.

Chapter 2 of the textbook provides information about linear, quadratic, and inverse relationships between variables.

Lab Partner™ Graphing Toolkit

A menu-driven graphing toolkit is available from the publisher to help plot graphs of experimental data. Graphs can be produced quickly, allowing time previously spent plotting data to be used for interpreting results. Student-collected data are entered, in column format, into a spreadsheet. A graph can be made by plotting any one or more columns of data against another. The axes are self-scaling to produce the best-size graph. With the software, you can either plot individual data points as a scatter graph or connect the points. A statistics treatment procedure in the software can determine the best curve fit for your data, using a linear relationship to a ninth-order equation. The statistics procedure provides the equation for your data, the intercepts, the slope, the data correlation coefficient, and the standard error of the curve fit. Another unique feature of the program allows a data column to be treated by any standard mathematical function, including a derivative and integral to establish a new column.

Data are easily saved to a disk. Samples of the printouts of Lab Partner™ are printed in the Teacher Annotated Edition of the *Laboratory Manual*. The program is set up for Epson®-compatible printers and can be adapted for most other printers. Included is a utility program that allows users to define their own printer code parameters. The Lab Partner™ program is available from the publisher for IBM®-compatible computers with a 1.2-m, 5-$\frac{1}{4}$-inch drive, a 3-$\frac{1}{2}$-inch drive, or a hard drive. A full-page graph is produced on 9 pin, 24 pin, or laser quality printers. A more basic version of the program is available for the Apple 11® series computers, which produces a smaller size plot.

Safety in the Laboratory

If you follow instructions exactly and understand the potential hazards of the equipment and the procedure used in an experiment, the physics laboratory is a safe place for learning and applying your knowledge. You must assume responsibility for the safety of yourself, your fellow students, and your teacher. Here are some safety rules to guide you in protecting yourself and others from injury and in maintaining a safe environment for learning.

1. The physics laboratory is to be used for serious work.

2. Never bring food, beverages, or make-up into the laboratory. Never taste anything in the laboratory. Never remove lab glassware from the laboratory, and never use this glassware for eating or drinking.

3. Do not perform experiments that are unauthorized. Always obtain your teacher's permission before beginning an activity.

4. Study your laboratory assignment before you come to the lab. If you are in doubt about any procedure, ask your teacher for help.

5. Keep work areas and the floor around you clean, dry, and free of clutter.

6. Use the safety equipment provided for you. Know the location of the fire extinguisher, safety shower, fire blanket, eyewash station, and first-aid kit.

7. Report any accident, injury, or incorrect procedure to your teacher at once.

8. Keep all materials away from open flames. When using any heating element, tie back long hair and loose clothing. If a fire should break out in the lab, or if your clothing should catch fire, smother it with a blanket or coat or use a fire extinguisher. NEVER RUN.

9. Handle toxic, combustible, or radioactive substances only under the direction of your teacher. If you spill acid or another corrosive chemical, wash it off with water immediately.

10. Place broken glass and solid substances in designated containers. Keep insoluble waste material out of the sink.

11. Use electrical equipment only under the supervision of your teacher. Be sure your teacher checks electric circuits before you activate them. Do not handle electric equipment with wet hands or when you are standing in damp areas.

12. When your investigation is completed, be sure to turn off the water and gas and disconnect electrical connections. Clean your work area. Return all materials and apparatus to their proper places. Wash your hands thoroughly after working in the laboratory.

Physics: Principles and Problems

First Aid in the Laboratory

Report all accidents, injuries, and spills to your teacher immediately.

You Must Know   safe laboratory techniques.

where and how to report an accident, injury, or spill.

the location of first-aid equipment, fire alarm, telephone, and school nurse's office.

Situation	Safe Response
Burns	Flush with cold water.
Cuts and bruises	Treat as directed by instructions included in your first-aid kit.
Electric shock	Provide person with fresh air; have person recline in a position such that the head is lower than the body; if necessary, provide artificial respiration.
Fainting or collapse	See Electric shock.
Fire	Turn off all flames and gas jets; wrap person in fire blanket; use fire extinguisher to put out fire. DO NOT use water to extinguish fire, as water may react with the burning substance and intensify the fire.
Foreign matter in eyes	Flush with plenty of water; use eye bath.
Poisoning	Note the suspected poisoning agent; contact your teacher for antidote; if necessary, call poison control center.
Severe bleeding	Apply pressure or a compress directly to the wound and get medical attention immediately.
Spills, general hydrogen acid burns	Wash area with plenty of water; use safety shower; use sodium carbonate, $NaHCO_3$ (baking soda).
Base burns	Use boric acid, H_3BO_3.

Safety Symbols

In this *Laboratory Manual* you will find several safety symbols that alert you to possible hazards and dangers in a laboratory activity. Be sure that you understand the meaning of each symbol before you begin an experiment. Take necessary precautions to avoid injury to yourself and others and to prevent damage to school property.

	DISPOSAL ALERT This symbol appears when care must be taken to dispose of materials properly.		**SKIN PROTECTION SAFETY** This symbol appears when use of caustic chemicals might irritate the skin or when contact with microorganisms might transmit infection.
	BIOLOGICAL HAZARD This symbol appears when there is danger involving bacteria, fungi, or protists.		**CLOTHING PROTECTION SAFETY** This symbol appears when substances used could stain or burn clothing.
	OPEN FLAME ALERT This symbol appears when use of an open flame could cause a fire or an explosion.		**FIRE SAFETY** This symbol appears when care should be taken around open flames.
	THERMAL SAFETY This symbol appears as a reminder to use caution when handling hot objects.		**EXPLOSION SAFETY** This symbol appears when the misuse of chemicals could cause an explosion.
	SHARP OBJECT SAFETY This symbol appears when a danger of cuts or punctures caused by the use of sharp objects exists.		**EYE SAFETY** This symbol appears when a danger to the eyes exists. Safety goggles should be worn when this symbol appears.
	FUME SAFETY This symbol appears when chemicals or chemical reactions could cause dangerous fumes.		**POISON SAFETY** This symbol appears when poisonous substances are used.
	ELECTRICAL SAFETY This symbol appears when care should be taken when using electrical equipment.		**CHEMICAL SAFETY** This symbol appears when chemicals used can cause burns or are poisonous if absorbed through the skin.

Common Physical Constants

Absolute zero = $-273.15°C = 0$ K

Acceleration due to gravity at sea level, (Washington, D.C.): $g = 9.80$ m/s^2

Atmospheric pressure (standard): 1 atm = 1.013×10^5 Pa = 760 mm Hg

Avogadro's number: $N_A = 6.02 \times 10^{23}$ mol^{-1}

Charge of one electron (elementary charge): $e = -1.602 \times 10^{-19}$ C

Coulomb's law constant: $K = 9.0 \times 10^9$ N·m^2/C^2

Gas constant: $R = 8.31$ J/mol·K

Gravitational constant: $G = 6.67 \times 10^{-11}$ N·m^2/kg^2

Heat of fusion of ice: 3.34×10^5 J/kg

Heat of vaporization of water: 2.26×10^6 J/kg

Mass of electron: $m_e = 9.1 \times 10^{-31}$ kg = 5.5×10^{-4} u

Mass of neutron: $m_n = 1.675 \times 10^{-27}$ kg = 1.00867 u

Mass of proton: $m_p = 1.673 \times 10^{-27}$ kg = 1.00728 u

Planck's constant: $h = 6.625 \times 10^{-34}$ J/Hz (J·s)

Velocity of light in a vacuum: $c = 2.99792458 \times 10^8$ m/s

Conversion Factors

Mass:	1000	g	= 1 kg
	1000	mg	= 1 g
Volume:	1000	mL	= 1 L
	1	mL	= 1 cm^3
Length:	1000	mm	= 1 m
	100	cm	= 1 m
	1000	m	= 1 km

1 atomic mass unit (u) = 1.66×10^{-27} kg = 931 MeV/c^2

1 electron volt (eV) = 1.602×10^{-19} J

1 joule (J) = 1 N·m = 1 V-C

1 coulomb = 6.242×10^{18} elementary charge units

Color Response of Eye to Various Wavelengths of Light

Color	Wavelength	
	in nm	in m
Ultraviolet	less than 380	less than 3.8×10^{-7}
Violet	400–420	$4.0–4.2 \times 10^{-7}$
Blue	440–480	$4.4–4.8 \times 10^{-7}$
Green	500–560	$5.0–5.6 \times 10^{-7}$
Yellow	580–600	$5.8–6.0 \times 10^{-7}$
Red	620–700	$6.2–7.0 \times 10^{-7}$
Infrared	above 760	above 7.6×10^{-7}

Prefixes Used with SI Units

Prefix	Symbol	Multiplier	Prefix	Symbol	Multiplier
Pico	p	10^{-12}	Tera	T	10^{12}
Nano	n	10^{-9}	Giga	G	10^{9}
Micro	μ	10^{-6}	Mega	M	10^{6}
Milli	m	10^{-5}	Kilo	k	10^{3}
Centi	c	10^{-2}	Hecto	h	10^{2}
Deci	d	10^{-1}	Deka	da	10^{1}

Properties of Common Substances

Specific Heat and Density		
Substance	**Specific heat (J/kg·K)**	**Density**
Alcohol	2450	0.8
Aluminum	903	2.7
Brass	376	8.5 varies by content
Carbon	710	1.7–3.5
Copper	385	8.9
Glass	664	2.2–2.6
Gold	129	19.3
Ice	2060	0.92
Iron (steel)	450	7.1–7.8
Lead	130	11.3
Mercury	138	13.6
Nickel	444	8.8
Platinum	133	21.4
Silver	235	10.5
Steam	2020	—
Tungsten	133	19.3
Water	4180	1.0 at 4°C, 0.99 at 0°C
Zinc	388	7.1

Index of Refraction	
Substance	**Index of refraction**
Air	1.00029
Alcohol	1.36
Benzene	1.50
Beryl	1.58
Carbon dioxide	1.00045
Cinnamon oil	1.6026
Clove oil	1.544
Diamond	2.42
Garnet	1.75
Glass, crown	1.52
Glass, flint	1.61
Mineral oil	1.48
Oil of wintergreen	1.48
Olive oil	1.47
Quartz. fused	1.46
Quartz, mineral	1.54
Topaz	1.62
Tourmaline	1.63
Turpentine	1.4721
Water	1.33
Water vapor	1.00025
Zircon	1.87

Spectral Lines of Elements

Element	Wavelength (nanometers)	Color
Argon	many close lines	
	706.7	red (strong)
	696.5	red (strong)
	603.2	orange
	591.2	orange
	588.8	orange
	550.6	yellow
	545.1	green
	525.2	green
	522.1	green
	518.7	green
	451.0	purple
	433.3	purple
	430.0	purple
	427.2	purple
	420.0	purple
Barium	659.5	red
	614.1	orange
	585.4	yellow
	577.7	yellow
	553.5	green (strong)
	455.4	blue (strong)
Bromine	many green lines	
	481.7	green
	478.6	blue
	470.5	blue
	many purple lines	
Calcium	445.4	blue
	443.4	blue-violet
	442.6	violet (strong)
	396.8	violet (strong)
	393.3	violet (strong)
Chromium	520.8	green
	520.6	green
	520.4	green
	428.9	violet (strong)
	427.4	violet (strong)
	425.4	violet (strong)
Copper	521.8	green
	515.3	green
	510.5	green
Hydrogen	656.2	red
	486.1	green
	434.0	blue-violet
	410.1	violet
Helium	706.5	red
	667.8	red
	587.5	orange (strong)
	501.5	green
	471.3	blue
	388.8	violet (strong)

Spectral Lines of Elements (continued)

Element	Wavelength (nanometers)	Color
Iodine	many lines	
	546.4	green (strong)
	516.1	green (strong)
Krypton	many faint close lines	
	587.0	orange (strong)
	557.0	yellow (strong)
	455.0	blue
	442.5	blue
	441.0	blue
	430.2	blue-violet
Mercury	623.4	red
	579.0	yellow (strong)
	576.9	yellow (strong)
	546.0	green (strong)
	435.8	blue-violet
	many lines in the violet and ultra violet	
Nitrogen	567.6	green (strong)
	566.6	green
	410.9	violet (strong)
	409.9	violet
Potassium	404.7	violet (strong)
	404.4	violet (strong)
Lithium	670.7	red (strong)
	610.3	orange
	460.3	violet
Sodium	589.5	yellow (strong)
	588.9	yellow (strong)
	568.8	green
	568.2	green
Neon	many lines in the red	
	640.2	orange (strong)
	585.2	yellow (strong)
	583.2	yellow (strong)
	540.0	green (strong)
Strontium	496.2	blue-green
	487.2	blue
	483.2	blue
	460.7	blue (strong)
	430.5	blue-violet
	421.5	violet
	407.7	violet
Xenon	492.3	blue-green
	484.4	blue
	482.9	blue
	480.7	blue
	469.7	blue
	467.1	blue (strong)
	462.4	blue (strong)
	460.3	blue
	458.3	blue
	452.4	blue
	450.0	blue (strong)

Rules for the Use of Meters

Introduction

Electric meters are precision instruments and must be handled with great care. They are easily damaged physically or electrically and are expensive to replace or repair. Meters are damaged physically by bumping or dropping and electrically by allowing excessive current to flow through the meter. The heating effect in a circuit increases with the square of the current. The wires inside the meter actually burn through if too much current flows through the meter. When possible, use a switch in the circuit to prevent having the circuit closed for extended periods of time. Note that meters are generally designed for use in either AC or DC circuits and are not interchangeable. In DC circuits, the polarity of the meter with respect to the power source in the circuit is critical. Be sure to use the proper meter for the type of circuit you are investigating. Always have your teacher check the circuit to be sure you have assembled it correctly.

The Voltmeter

A voltmeter is used to determine the potential difference between two points in a circuit. It is always connected in parallel, never in series, with the element to be measured. If you can remove the voltmeter from the circuit without interrupting the circuit, you have connected it correctly.

On a DC voltmeter, the terminals are marked $+$ or $-$. The positive terminal should be connected either directly or through components to the positive side of the power supply. The negative terminal must be connected either directly or through circuit components to the negative side of the power supply. After you have connected the voltmeter, close the switch for a moment to see if the polarity is correct.

Some meters have several ranges from which to choose. You may have a meter with ranges from 0–3 V, 0–15 V, and 0–300 V. If you do not know the potential difference across the circuit on which the voltmeter is to be used, choose the highest range initially, and then adjust to a range that gives readings in the middle of the scale (when possible).

The Ammeter

An ammeter is used to measure the current in a circuit and must always be connected in series. Since the internal resistance of an ammeter is very small, the meter will be destroyed if it is connected in parallel. When you connect or disconnect the ammeter, the circuit must be interrupted. If the ammeter can be included or removed without breaking the circuit, the ammeter is incorrectly connected.

Like a voltmeter, an ammeter may have different ranges. Always protect the instrument by connecting it first to the highest range and then proceeding to a smaller scale until you obtain a reading in the middle of the scale (when possible).

On a DC ammeter, the polarity of the terminals is marked $+$ or $-$. The positive terminal should be connected either directly or through components to the positive side of the power supply. The negative terminal must be connected either directly or through circuit components to the negative side of the power supply. After you have connected the ammeter, close the switch for a moment to see if the polarity is correct.

The Galvanometer

A galvanometer is a very low resistance instrument used to measure very small currents in microamperes. Thus, it must be connected in series in a circuit. The zero point on some galvanometers is in the center of the scale, and the divisions are not calibrated. This type of galvanometer measures the presence of a very small current, its direction, and its relative magnitude. A low resistance wire, called a shunt, may be connected across the terminals of the galvanometer to protect it. If the meter does not register a current, the shunt is then removed.

Potentiometer (variable resistors)

A potentiometer is a precision instrument that contains an adjustable resistance element. When placed in a circuit, a potentiometer allows gradual changing of the resistance which, in turn, causes the current and voltage to vary. Some power resistors are called rheostats.

Resistor Color Code

Suppose that a resistor has the following four color bands:

1st band	2nd band	3rd band	4th band
brown	black	yellow	gold
1	0	4	± 5%

The value of this resistor is $10 \times 10\ 000 \pm 5\% = 100\ 000 \pm 5\%$ or it has a range of $80\ 000\ \Omega$ to $120\ 000\ \Omega$.

Resistance Color Codes (resistance given in ohms)			
Color	**Digit**	**Multiplier**	**Tolerance (%)**
Black	0	1	
Brown	1	10	
Red	2	100	
Orange	3	1000	
Yellow	4	10 000	
Green	5	100 000	
Blue	6	1 000 000	
Violet	7	10 000 000	
Gray	8		
White	9		
Gold		0.1	5
Silver		0.01	10
No color			20

Preparation of Formal Laboratory Reports

Ordinarily, the reports you write for the experiments in this manual will be simple summaries of your work. Laboratory data sheets are provided in the manual for this purpose. In the future, you may be called upon to write more formal reports for other science courses. There is an accepted procedure for writing these reports. The procedure is outlined below. Your teacher may require that you write a number of reports in accordance with this outline so that you learn how to prepare reports properly.

Sections to be included in a formal laboratory report include the following: Introduction, Data, Results/Analysis, Graphs, Sample Calculations, Discussion, Conclusions.

I. Introduction

A. Heading

This includes the experiment number and title, date, name, and your partner's name if you do a joint experiment. When two students work together using the same apparatus, they are partners for data collection purposes, yet each must write a separate report.

B. Diagrams

1. Make sketches of mechanical apparatus (if called for).

2. Draw complete electric circuit diagrams showing all electrical components. Label the polarity of all DC meters and power sources.

C. Provide a brief explanation or title for each diagram.

D. Include an elaboration or a summary of the concept, purpose, procedure, theory, or the history of the experiment.

II. Data

A. Use only the original record of the measurements made during the experiment. Never jot the data down on scrap paper for future use. Prepare a data sheet and use it.

III. Results/Analysis

A. The results section consists of a tabulation of all intermediate calculated values and final results.

B. Whenever there are several results, the numerical values should be recorded in a table.

C. Tables must have titles. Headings and extra notes may be required to make the analysis or significance of the results clear to the reader.

IV. Graphs

A. Use adequate labels (title, legend, names of quantities and units).

B. Draw the best, smooth curve possible; do not draw curves dot-to-dot.

V. Sample Calculations

A. Each sample calculation should include the following items:

1. an equation in a familiar form

2. an algebraic solution of the equation for the desired quantity

3. substitution of known values with units

4. numerical answer with units

For example, if $d = 10$ m and $t = 2$ s, to solve for a:

using $d = v_i t + \frac{1}{2} a t^2$, where $v_i = 0$,

$$a = 2d/t^2 = (2)(10 \text{ m})/(2 \text{ s})^2 = 5 \text{ m/s}^2.$$

VI. Discussion

In some cases, the conclusions of an experiment are so obvious that the discussion section may be omitted. However, in these instances a short statement is appropriately included. More often, some discussion of the results will be required to make their significance clear. You may also wish to comment upon possible sources of error and to suggest improvements in the procedure or apparatus.

VII. Conclusions

The conclusion is an important part of every report. The conclusion must be the individual work of the student who writes the report and should be completed without the assistance of anyone, unless it is the teacher.

The conclusion consists of one or more well-written paragraphs summarizing and drawing together only the main results and indicating their significance in relationship to the observed data.

A. Conclusions must cover each point of the subject.

B. Conclusions must be based upon the results of the experiment and the data.

C. If conclusions are based upon graphs, reference must be made to the graph by its full title.

D. Clarity and conciseness are particularly important in conclusions. The personal form should be avoided except, perhaps, in the discussion. Therefore, do not use the words *I* or *we* unless there is a special reason for doing so.

1-1

Physics Lab

Bubble Up

When you do the activities in this manual, you will acquire data by making observations. The observations will be either qualitative, such as color or shape, or quantitative, such as a measurement of length, mass, or rate of activity. The quality of your observations and how you organize and communicate data can affect your ability to draw inferences, form hypotheses, and predict outcomes.

Procedure

1. Soak two raisins overnight in a 100-mL beaker filled with tap water. **CAUTION:** *Eating and drinking in the school science laboratory is never safe. Do not eat the raisins or drink the liquids used in this experiment.*

2. Pour about 100 mL of carbonated water into a 150-mL beaker and observe the water closely for 1 min. Record your observations in Table 1.

3. Using a spoon, remove one raisin from the tap water. In Table 1, describe the raisin. Compare the appearance of the soaked raisin to that of a dry, normal raisin. Use a metric ruler to measure each type of raisin. Record your observations in Table 1.

4. Place the soaked raisin in the carbonated water. Watch closely for the next 5 min and record your observations in Table 2.

Data and Observations

Table 1	
Material	**Observations**
Carbonated water	
Soaked raisin	
Dry raisin	

Objectives

- **Observe** similarities and differences between carbonated and tap water in their effect on the motion of a raisin.
- **Form** hypotheses about your observations.
- **Design** an experiment to test a hypothesis.
- **Communicate** an experimental plan and results.

Materials

- 100-mL beaker
- 150-mL beaker
- 100 mL carbonated water
- 4 raisins
- clock
- 100 mL tap water
- plastic spoon
- metric ruler

1-1 Physics Lab

Table 2
Observations of soaked raisin in carbonated water

Analysis and Conclusions

1. Which of your observations are not related to the motion of the raisin?

_____

_____

2. Which of your observations are related to the motion of the raisin?

_____

_____

3. Form a hypothesis, or generalization, to explain the observed motion of the raisin.

_____

_____

_____

4. Using the materials listed for this lab, design an experiment to test your hypothesis. Write down your procedure. Identify which variables will remain constant and which will be manipulated. Predict a way to stop the motion of the raisin.

_____

_____

_____

_____

_____

_____

_____

_____

1-1 Physics Lab

• • • • • • • • • • • • • •

5. Describe your experiment to your teacher. With the teacher's permission, perform your experiment. Was your prediction correct? Did the outcome support or refute your hypothesis?

_____

_____

_____

Extension and Application

1. Why do you exhale before you do a surface dive in a lake or a swimming pool? Why do some people float better than others?

_____

_____

_____

_____

2. Why is air pumped into a sunken ship to raise it to the surface?

_____

_____

_____

3. Use the library or the Internet to find out how submarines sink or rise in the ocean.

_____

_____

_____

_____

_____

2-1

Physics Lab

Measuring Length

The only certain thing about measurements in experiments is that there is no certainty. The best you can do is minimize the uncertainty in an experiment. The degree of uncertainty depends on the precision of the measuring instrument and the skill of the user. Instruments with more divisions are usually more precise and have a precision equal to one-half the smallest division of the instrument. For example, the precision of a meterstick is 0.5 mm, or 0.005 (5×10^{-3}) m.

You can be a skillful user of a meterstick by following these basic rules. Read the measurements from directly in front of the meterstick at eye level. Be consistent in technique. Don't use the markings at the end of a meterstick. Use a mark farther along the stick and subtract the lower measure from the higher one, as shown in Figure A.

Objectives

- **Measure** the dimensions of several objects, using SI units.
- **Calculate** the volume of a rectangular solid.
- **Calculate** the thickness of a single page of the Lab Manual.
- **Apply** significant digits to measurements and calculations.

Procedure

1. Measure and record in Table 1 the length, width, and height in centimeters of a block of wood. Measure to the nearest millimeter, and then estimate to a tenth of a millimeter. The estimated number and all figures to the left of it are significant digits.

7.35 cm − 5.00 cm = 2.35 cm

Figure A

2. Measure the length, width, and height in centimeters of your Laboratory Manual. Record the measurements in Table 1.

3. Calculate the volume of the block in cubic centimeters. Include only as many significant digits as you have in the value for one dimension of the block. Record the volume in Table 1.

4. Calculate the volume of the Laboratory Manual in cubic centimeters. Include only as many significant digits as you have in the value for one dimension of the manual. Record the volume in Table 1.

5. Place the meterstick next to the edge of one page of your Laboratory Manual. Estimate the thickness of one page in millimeters. Record your estimate in Table 2.

6. Measure the thickness of 50, 75, and 100 pages of your Laboratory Manual in millimeters. Record your results in Table 2. For each quantity of pages, determine the average thickness of a single page. Record your calculations in the table.

Materials

- meterstick
- wood block
- Laboratory Manual

2-1 Physics Lab

....

Data and Observations

Table 1				
	Length (cm)	**Width (cm)**	**Height (cm)**	**Volume (cm³)**
Wood block				
Laboratory Manual				

Table 2		
Number of pages	**Total thickness (mm)**	**Average thickness of a single page (mm)**
1		
50		
75		
100		

2-1 Physics Lab

Analysis and Conclusions

1. Why is it less accurate to use either end of the meterstick for measuring length?

   _____

   _____

2. Of all the measurements you made, which one do you feel is least precise? Give a reason for your answer.

   _____

   _____

   _____

3. If the values in Table 2 for the average thickness of a single page are different, which value is probably the most accurate? Give a reason for your answer.

   _____

   _____

   _____

   _____

Extension and Application

1. How important are measuring instruments with a high degree of precision to a craftsperson manufacturing picture frames?

   _____

   _____

   _____

   _____

   _____

2-2

Design Your Own **Physics Lab**

Measuring Temperature

Objectives

- **Measure** the temperatures of several samples of water.
- **Compare** the values obtained.
- **Infer** the effects of measurement on the reliability of results.

Getting valid measurements is not always easy. Many things interfere with accuracy and precision. Most obviously, measuring devices can be faulty or incorrectly calibrated, or the person taking the readings can make a mistake. The act of measuring itself can introduce uncertainty. The measurer and the measuring device interact with the system being measured and affect the system and the resulting data.

In this lab, you will measure the temperature of several volumes of water. The water will come from the same source under the same initial conditions, so the temperature must initially be identical for all samples. You'll see how taking samples and using a thermometer on them affects the readings you get.

Problem

How do the quantity of a sample, the elapsed time, and the equipment affect the reliability of a measurement?

Hypothesis

Formulate a hypothesis about how variables in an experiment affect your results.

_____

_____

Possible Materials

CBL unit

temperature probe

graphing calculator

link cable

4 beakers

3 graduated cylinders

tap water

4 thermometers

safety goggles

heat-protective gloves

tongs

electric hot plate

stopwatch

Plan the Experiment

1. Work with a partner or in a small group. Decide on a procedure that uses the suggested materials (or others of your choosing) to measure the temperature of three different volumes of water by two different techniques: a) using standard thermometers, and b) using a CBL unit. All the samples are to be drawn from a single large source that has been heated to a temperature well above room temperature. The suggested volumes of the three samples are 80 mL, 5 mL, and 1 mL.

2. Decide what kind of data to collect and how to analyze it. You can record your data in the table on the next page. Label the columns appropriately.

3. Write your procedure on another sheet of paper or in your notebook.

4. **Check the Plan** Have your teacher approve your plan before you proceed with your experiment. Be careful with the hot plate and heated water. Do not place the CBL temperature probe into boiling water.

2-2 Physics Lab

Data and Observations

Room temperature: _____ °C

Temperature of initial large sample of heated water: _____ °C

Data Table			

Analyze and Conclude

1. **Graphing Data** Use the grid on the next page to graph temperature versus time for the three samples over the measurement period. What do the graphs have in common? How do they differ? Compare the temperatures of the samples when you last measured them. Were the temperatures the same? How did they compare?

_____

_____

_____

_____

_____

_____

_____

_____

_____

_____

2-2 Physics Lab

Temperature

Time

2. **Observing and Inferring** Account for any measured temperature differences among the three samples of water. Compare the results for parts A and B. What might account for this result?

_____

_____

_____

_____

3. **Using Graphs** Extrapolate (extend) the graphed lines to the y-axis. Do any of the y-intercepts correspond to the temperature of the original large quantity of heated water? Would you expect them to? Give reasons for your answer.

_____

_____

_____

_____

4. **Predicting** To what temperature would the water in the samples have eventually fallen? Give a reason for your answer.

_____

_____

2-2 Physics Lab

· · · · · · · · · · · · · ·

5. **Checking Your Hypothesis** You're by the cold North Atlantic on a warm August day. If you scooped a small tube of water from the sea and measured its temperature and that of the sea simultaneously, would you get the same readings? How about 5 min later? Give reasons for your answers.

_____

_____

_____

_____

_____

6. **Interpreting** Apply the following statement to what you learned in this investigation: The act of observing a system affects the system.

_____

_____

_____

_____

_____

Apply

1. A radar gun used to measure the speeds of moving cars transmits radio waves. The waves strike the cars for a short time and are partially reflected to a receiver in the radar gun. The device interprets the differences in the reflection time of the radio waves, thereby measuring the car's speed. Does the waves' effect on the system being measured affect the reliability of the measurement? Give a reason for your answer.

_____

_____

_____

_____

3-1

Physics Lab

Analyzing Motion

In this activity, you will be investigating the motion of a moving vehicle to determine its velocity. The CBL unit with a motion detector will measure time intervals and the vehicle's displacement. The ultrasonic motion detector emits short pulses of ultrasonic sound waves and then listens for a reflected echo of the pulse. The CBL unit uses the speed of the ultrasonic sound waves and the time it took to return to the detector to determine the distance to the object from which the waves were reflected.

Objectives

- **Calculate** the average velocity for a series of intervals.
- **Classify** the displacement: time ratio as uniform or nonuniform motion.
- **Analyze** the relationship between total displacement and time, and velocity and time.
- **Define** constant velocity operationally.
- **Interpret** data to establish trends.

Constant-velocity vehicle

Vehicle motion

Ultrasonic motion detector

Figure A

Materials

- constant-velocity vehicle
- metric ruler
- C-clamp
- CBL unit
- graphing calculator
- link cable
- ultrasonic motion detector
- graph paper

The CBL unit will collect data of time and associated distances of the moving vehicle from the detector. The average velocity can be determined by

$$\bar{v} = \frac{\Delta d}{\Delta t} = \frac{d_1 - d_0}{t_1 - t_0}$$

Procedure

1. With your lab partner, set up the ultrasonic motion detector and vehicle as shown in Figure A. Place the vehicle about 0.2 m from the motion detector.

2. Connect the CBL unit to the calculator using the unit-to-unit link cable and the link ports on the calculator and CBL unit. Connect the ultrasonic motion detector to the SONIC port of the CBL unit.

3-1 Physics Lab

3. Turn on the CBL unit and graphing calculator. If not already loaded into your calculator, load the program PHYSICS and its associated programs into your calculator from another calculator or download them from a computer. Start the program PHYSICS on your graphing calculator.

4. Select the option SET UP PROBES from the MAIN MENU. Enter 1 for the number of probes. Then from the SELECT PROBE menu, choose MOTION from the list.

5. Select the MONITOR INPUT option from the COLLECT DATA menu. This will allow you to check and see if the motion detector is working properly, displaying the distance between the motion detector and the vehicle on the CBL unit. Press the "+" to quit the sampling test.

6. You are ready to begin the experiment. Select the COLLECT DATA option. In DATA COLLECTION, select TIME GRAPH. The calculator will prompt you to ENTER TIME BETWEEN SAMPLES IN SEC-ONDS. Enter 0.1 for the time between samples. Enter 99 for the number of samples. (A TI-82 can collect only 99, a TI-85 can collect only 55, while a TI-83 can collect 120.) Check the values you entered and press ENTER. If the values are correct, select USE TIME SETUP to continue. If you have made a mistake entering the values, then select MODIFY SETUP and reenter the values before con-tinuing.

7. On the TIME GRAPH menu, select NON-LIVE DISPLAY.

8. Press ENTER. The READY EQUIPMENT command should appear. One student should press ENTER on the graphing calculator. When the motion detector begins to click, another student should turn on the vehicle and release it so it moves across the table away from the motion detector.

9. When the motion detector has stopped clicking and the CBL unit displays DONE, press ENTER on the graphing calculator. From the SELECT GRAPH menu, select DISTANCE to plot a graph of the distance in meters against the time in seconds.

10. Use the arrow keys on the calculator to trace along the curve. On the far left, the curve represents the position of the vehicle before its motion began. The middle section of the curve represents the motion of the moving vehicle. Copy this graph into your laboratory report.

11. Chose a point on the curve at the beginning of the middle section of the graph. Record the dis-tance (y-value) and the time (x-value) into your data table. Press the right arrow key again three times to move to another point along the curve. Record the new time and distance values in your data table. Continue moving along the curve in increments of three steps and recording data until ten sets of data have been recorded. Press ENTER to return to the SELECT GRAPH menu.

12. Select VELOCITY to plot a graph of velocity in m/s against time in seconds. Observe how the velocity of the vehicle changed during the experiment. Sketch this graph in your laboratory report. Press ENTER.

13. Select ACCELERATION to plot a graph of the acceleration in m/s^2 against the time in seconds. Observe how the acceleration of the vehicle changed during the experiment. Sketch this graph in your laboratory report. Press ENTER. Turn off your CBL unit and calculator.

3-1 Physics Lab

Data and Observations

Time (s)	Displacement (m)	Total displacement (m)	Average velocity m/s

Analysis and Conclusions

1. Using your distance verses time graph for the vehicle, describe the motion.

   _____

   _____

   _____

2. Using Table 1, calculate the average velocity between each set of displacements.

3. Describe the acceleration verses time graph for the middle portion of the vehicle's motion.

   _____

   _____

   _____

Name _____

4. Are there any time intervals on the curve during which the velocity of the vehicle appears to be changing? How do you recognize those intervals?

_____

_____

_____

5. Are there any time intervals on the curve in which the velocity of the vehicle appears to be constant?

_____

_____

_____

6. Calculate the vehicle's average velocity for the entire trip. How does it compare to the velocities during each interval?

_____

_____

Extension and Application

1. Using the data, calculate the average acceleration for your vehicle.

2. Suppose that a constant-velocity vehicle could double its average velocity. How would this change affect

   a. the distance in a given interval of time?

   _____

   b. the overall displacement of the vehicle in the same time?

   _____

   c. the acceleration of the vehicle?

   _____

4-1

Physics Lab

Addition of Vector Forces

When two or more forces act at the same time on an object and their vector sum is zero, the object is in equilibrium. Each of the arrangements shown in Figure A illustrates three concurrent forces acting on point P. Because point P is not moving, the three forces produce no net force on point P, and the systems are in equilibrium. In this experiment, you will determine the vector sum of two of the concurrent forces, called the resultant, and investigate the relationship of the resultant to the third force.

Objectives

- **Observe** the interaction of concurrent forces.
- **Apply** vector addition to obtain the resultant of forces in equilibrium.
- **Demonstrate** equivalent ways to add vectors.

Procedure

Your teacher will demonstrate the apparatus you will be using—either Method 1 or Method 2.

Materials

Method 1—force table and 3 spring scales

Method 2—2 spring scales, 2 ring stands, cross support, and 500-g mass

metric ruler

heavy string

pencil

protractor

paper

Figure A

You can use a variety of methods to measure forces. Your apparatus will probably be similar to one of these.

Method 1

1. Set up the apparatus shown in Figure A1. Each spring scale should read zero when no load is attached. Attach the spring scales to the force table so that each scale registers a force at approximately midrange.

2. Place a piece of paper beneath the spring scales. Using a sharp pencil, mark several points along the line of action (the string) of each force.

3. Remove the paper and, using the points that you marked, construct lines A, B, and C, representing the direction of force action for scales A, B, and C.

4-1 Physics Lab

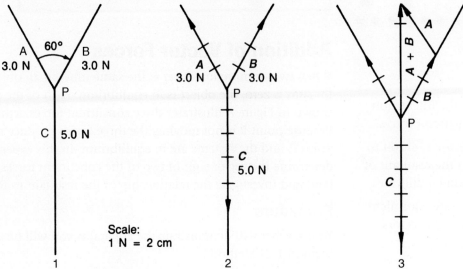

Figure B

Scale:
1 N = 2 cm

4. Record the reading of each spring scale next to its corresponding line, as shown in Figure B1.

5. Select a suitable scale, such as 1 N = 2 cm, and record your scale near line C. Construct vectors, of proper scaled length, along lines A, B, and C to represent each force. If the spring scales are not calibrated in newtons, convert the measurements to newtons by multiplying the mass in kilograms by 9.8 m/s². Figure B2 shows how to scale these vectors.

6. Add vector **A** to vector **B** by reproducing **A** parallel to itself but with its tail at the head of **B** (the head-to-tail method), as shown in Figure B3.

7. Draw a vector representing the vector sum of **A** + **B**, the resultant.

8. Repeat steps 1–7 so that each lab partner has a set of data to analyze.

Method 2

1. Set up the apparatus shown in Figure A2. With a protractor, measure each of the three angles at the intersection of the three strings.

2. Using these angle measurements, construct a diagram on paper of the forces acting on point P by drawing three lines to represent the lines of action of the three forces. Label the lines A, B, and C, as shown in Figure B1.

3. Record the values of the two spring-scale readings and the weight in newtons of the 500-g mass next to lines A, B, and C on your paper. If the scales measure mass, convert the mass readings to newtons by multiplying the mass in kilograms by 9.8 m/s².

4. Select a suitable scale, such as 1 N = 2 cm, and record your scale near line C. Using your scale, construct vectors along lines A, B, and C to represent the forces acting along each line of force, as shown in Figure B2.

5. Add vector **A** to vector **B** by reproducing **A** parallel to itself but with its tail at the head of **B** (the head-to-tail method), as shown in Figure B3.

6. Draw a vector representing the vector sum **A** + **B**, the resultant.

7. Repeat steps 1–6 so that each lab partner has a set of data to analyze.

Copyright © by Glencoe/McGraw-Hill

4-1 Physics Lab

Data and Observations

On a separate sheet of paper, draw the vectors from your experiment. Follow the format in Figure B.

Analysis and Conclusions

1. What scale did you select for your model? Compute the resultant using your number scale.

2. Compare the magnitude and direction of the computed resultant force of *A* + *B* with the measured or known magnitude of force *C*. Explain your findings. Calculate the relative error in the magnitudes using force *C* as the reference value.

3. On a separate street of paper, reconstruct vectors *A*, *B*, and *C* and add them. Do this by placing a piece of paper over your first diagram and tracing the vectors. Place the tail of *B* at the head of *A*, and then place the tail of *C* at the head of *B* (the head-to-tail method). Label your vectors.

4. Explain the results of the graphical addition of *A* + *B* + *C* from question 3.

   _____

   _____

   _____

   _____

5. Suppose that you had added *B* to *C*. What result would you expect?

   _____

   _____

   _____

6. What result would you expect if you added *C* to *A*?

   _____

   _____

   _____

7. On the same paper you used for question 3, add your three vectors in the order *C* + *B* + *A*. What is the resultant?

   _____

4-1 Physics Lab

Extension and Application

1. Using a different method, repeat the investigation, setting the angle between A and B at an angle other than 90°. Solve for C both mathematically and graphically.

2. A sky diver jumps from a plane and, after falling for 11 s, reaches a terminal velocity (constant speed) of 250 km/h. Changing her body configuration, she accelerates to a terminal velocity of 320 km/h. Finally, after releasing her parachute, she again accelerates and reaches a final terminal velocity of 15 km/h. Explain how it is possible to obtain the different terminal velocities.

_____

_____

_____

_____

_____

_____

_____

_____

5-1

Physics Lab

Accelerated Motion

The ultrasonic motion detector can record the movement of a small cart pulled across a table by a falling mass. The resulting data measures the displacement of the moving cart per time. From Lab 3-1, you know that average velocity equals displacement for a given interval of time. Also recall that the ratio of a change in velocity to a change in time is acceleration,

$$a = \frac{v_2 - v_1}{t_2 - t_1},$$

and that this is the equation for the slope of a graph of velocity versus time.

Objectives

- **Measure** the displacement of a moving object in a set time interval.
- **Calculate** the velocity of a moving object.
- **Analyze** motion, using graphs of the relationship between displacement and time, velocity and time, and acceleration and time.

Procedure

1. Set up the apparatus as shown in Figure A, except for the 500-g mass. Put the motion detector about 1 m from the pulley. Tie 1 m of string to the opposite end of the cart and thread the string through the pulley.

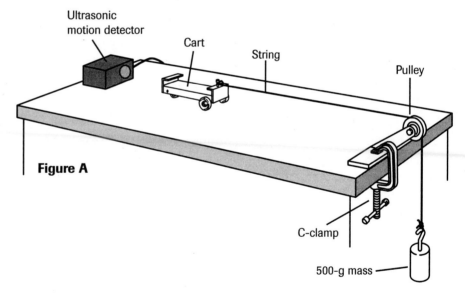

Figure A

Ultrasonic motion detector · Cart · String · Pulley · C-clamp · 500-g mass

Materials

- laboratory cart
- CBL unit
- ultrasonic motion detector
- link cable
- graphing calculator
- 500-g mass
- masking tape
- heavy string, 1 m
- pulley
- C-clamp
- graph paper

2. With your lab partner, set up the ultrasonic motion detector and vehicle as shown in Figure A. Place the motion detector about 0.4 m from the cart.

3. Connect the CBL unit to the calculator using the unit-to-unit link cable and the link ports on the calculator and CBL unit. Connect the ultrasonic motion detector to the SONIC port of the CBL unit.

4. Turn on the CBL unit and graphing calculator. If not already loaded into your calculator, load the program PHYSICS and its subprograms into your calculator from another calculator or download them from a computer. Start the program PHYSICS on your graphing calculator.

5-1 Physics Lab

5. Select the option SET UP PROBES from the MAIN MENU. Enter 1 for the number of probes. Then from the SELECT PROBE menu, choose MOTION from the list.

6. Select the MONITOR INPUT option from the COLLECT DATA menu. This will allow you to check and see if the motion detector is working properly by displaying the distance between the motion detector and the vehicle on the CBL unit. Press the "+" to quit the sampling test.

7. You are ready to begin the experiment. Select the COLLECT DATA option. In DATA COLLECTION, select TIME GRAPH. The calculator will prompt you to ENTER TIME BETWEEN SAMPLES IN SECONDS. Enter 0.02 for the time between samples. Enter 99 for the number of samples. (A TI-82 can collect only 99, a TI-85 can collect only 55, while a TI-83 can collect 120.) Check the values you entered and press ENTER. If the values are correct, select USE TIME SETUP to continue. If you have made a mistake entering the values, then select MODIFY SETUP and reenter the values before continuing.

8. On the TIME GRAPH menu, select NON-LIVE DISPLAY.

9. Press ENTER. The READY EQUIPMENT command should appear. While preventing the cart from moving, one student should press ENTER on the graphing calculator. When the motion detector begins to click, another student should release the cart, allowing the 500-g mass to pull the cart across the table. Catch the cart at the edge of the table to prevent it from knocking the pulley loose or plunging to the floor.

10. When the motion detector has stopped clicking and the CBL unit displays DONE, press ENTER on the graphing calculator. From the SELECT GRAPH menu, select DISTANCE to plot a graph of the distance in meters against the time in seconds.

11. The displacement graph will initially show a horizontal line until the cart begins to move. Copy this graph into your laboratory report.

12. Using the right arrow key, move along the curve until the cart begins to move and then back up one point. Record the distance (y-value) and the time (x-value) into your data table. Press the right arrow key again four times to move to another point along the curve. Record the new time and distance values in your data table. Continue moving along the curve in increments of four steps and recording data until ten sets of data have been recorded. Press ENTER to return to the SELECT GRAPH menu.

13. Select VELOCITY to plot a graph of velocity in m/s against time in seconds. Observe how the velocity of the vehicle changed during the experiment. Sketch this graph in your laboratory report. Press ENTER.

14. Select ACCELERATION to plot a graph of the acceleration in m/s^2 against the time in seconds. Observe how the acceleration of the vehicle changed during the experiment. Sketch this graph in your laboratory report. Press ENTER. Turn off your CBL unit and calculator.

5-1 Physics Lab

Data and Observations

Table 1	
Time (s)	Total displacement (m)

Table 2	
Time (s)	Average velocity (m/s)

5-1 Physics Lab

Analysis and Conclusions

1. On graph paper, plot the total displacement of the cart versus the time. Use the values from Table 1.

2. On your velocity graph, use a colored pencil to make a smooth best-fit curve.

3. Describe the graph of displacement versus time. What does the graph mean?

   _____

   _____

   _____

   _____

   _____

4. Describe the graph of velocity versus time. What does the graph mean?

   _____

   _____

   _____

   _____

5. Calculate the average velocity and enter it in Table 2. Using your data for average velocity, calculate the acceleration, $\Delta v / \Delta t$, for each interval. Because v_0 is zero, the change in velocity is equal to the average velocity in each interval. You can find the acceleration by dividing each change in velocity by the time.

6. On graph paper, plot acceleration versus time. Describe this graph. What does your graph show about the acceleration of the cart? How does this graph compare to your calculator?

   _____

   _____

   _____

5-1 Physics Lab

Extension and Application

1. At several points along the graph of total displacement versus time, draw a tangent to the curve and find the slope. Compare the slopes to the average speeds for the various time intervals.

2. Compute your acceleration in meters per seconds squared. How does your cart's acceleration compare to the acceleration due to gravity?

_____

_____

3. What would be the advantage of building an interplanetary spacecraft that could accelerate at 1 g (9.8 m/s^2) for a year?

_____

_____

_____

_____

_____

_____

5-2

Design Your Own Physics Lab

Acceleration Due to Gravity

Objectives

- **Measure** displacements and times for a falling mass.
- **Calculate** average velocity.
- **Determine** acceleration by analyzing data for velocity and time.

Acceleration is the rate of change of velocity. To determine acceleration, calculate the average velocity, $\bar{v}$, for each time interval, by dividing the displacement by the time interval.

$$\bar{v} = \frac{\Delta d}{\Delta t}$$

If the average velocity increases or decreases from interval to interval, there is acceleration, a. The acceleration is equal to the velocity change per unit time.

$$a = \frac{\Delta v}{\Delta t}$$

If an object accelerates uniformly, the value of a is constant. A plot of velocity versus time for uniform acceleration is a straight line. The slope of the line is equal to the acceleration. The slope is the ratio of rise to run. In this case, the rise is equal to the difference between the final velocity and the initial velocity ($v_f - v_i$). The run is the difference between the final time and the initial time ($t_f - t_i$).

$$a = \text{slope} = \frac{v_f - v_i}{t_f - t_i}$$

For an object dropped from rest, the relationship between the object's displacement, d, and elapsed time, t, is the following.

$$d = \frac{1}{2} gt^2$$

In the equation, g is acceleration due to gravity. If you measure displacement and elapsed time, you can calculate the value of g by rearranging the equation.

$$g = 2d/t^2$$

You could use a recording timer to determine the values of d and t. This device puts dots on a tape. The period, or time interval between dot markings, is fixed and will be given to you by your teacher. If the tape is attached to a moving object, the tape moves through the timer and dots appear along the tape. You can label the dots, starting with a zero, so the first interval is between dot 0 and dot 1, the second interval is between dots 1 and 2, and so on. The distances between dots give you the displacement. The total elapsed time from dot 0 to any other dot equals the interval number times the period of the timer.

5-2 Physics Lab

Possible Materials

recording timer with
power supply

timer tape

carbon-paper disks

C-clamp

masking tape

1-kg mass

centimeter ruler

Problem

How can the acceleration due to gravity be determined by measuring displacements and times?

Hypothesis

Formulate a hypothesis that would describe how to obtain displacement and time information for use in calculating a value for g that is close to the accepted value.

_____

_____

Plan the Experiment

1. Work in pairs or small groups. Decide on a procedure that uses the suggested materials (or others of your choosing) to measure a number of displacement intervals and falling times for the 1-kg mass.

2. Decide what kind of data to collect and how to analyze it to get a value for the acceleration due to gravity. Use the space on the next page to create a table to record your data and calculated results.

3. Write your procedure on another sheet of paper or in your notebook. In the space below, draw the setup you plan to use.

4. **Check the Plan** Have your teacher approve your plan before you proceed with your experiment. Make sure that you understand how to operate the recording timer. Be careful with the electric power source.

Setup

5-2 Physics Lab

Data and Observations

Period of recording timer: _____ s

	Table 1			
	Time, Displacement, and Velocity Data			
Interval	Total time (s)	Displacement in interval (m)	Total displacement (m)	Average velocity (m/s)
1				
2				
3				
4				
5				
6				
7				
8				
9				
10				
11				
12				
13				
14				
15				
16				
17				
18				
19				
20				

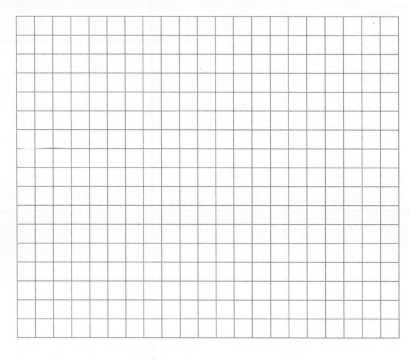

5-2 Physics Lab

Analyze and Conclude

1. **Calculating Results** Based on your timer tape and the period of the recording timer, determine values for each interval for the total time elapsed, the displacement during each interval, and the total displacement. Find the average velocity for each interval.

2. **Analyzing Results** What does the relationship between total displacement and total time indicate about the motion of the falling mass? Use the grid below to graph your results.

3. **Analyzing Results** What does the relationship between average velocity and total time indicate about the motion of the falling mass? Use the grid below to graph your results.

Total Displacement

Average Velocity

Total Time

4. **Calculating Results** Use your results to find a value for g. Compare this value to the accepted value of 9.80 m/s^2.

_____

Apply

1. How would the value of g measured at a higher elevation compare to the value you measured? Would this difference affect athletes in Olympic Games at high altitudes? Give a reason for your answer. What other factor might offset any advantage or disadvantage?

_____

_____

_____

_____

6-1

Physics Lab

Newton's Second Law

Newton's second law of motion states that the acceleration of a body is directly proportional to the net force on it and inversely proportional to its mass. In mathematical form, this law is expressed as $a = F/m$. An object will accelerate if it has a net force acting on it, and the acceleration will be in the direction of the force. In this experiment, a laboratory cart will be accelerated by a known force, and its acceleration will be measured using a vernier motion detector. The product of the total mass accelerated and its acceleration equals the force causing the acceleration.

In order to calculate the net force acting on the laboratory cart, friction forces that oppose the motion must be offset. If the cart moves at a constant velocity, then the net force acting on the cart is zero because the acceleration is zero. The friction force can be neutralized by providing enough small masses at the end of the string to make the cart move forward at a constant velocity.

If you assume that the laboratory cart experiences a uniform acceleration due to the falling mass, then the following relationship applies.

$$d = v_i t + \frac{1}{2}at^2$$

If the cart has an initial velocity of zero, the equation simplifies as follows.

$$d = \frac{1}{2}at^2$$

If the displacement in a given time is known, you can solve for acceleration using this form of the equation.

$$a = 2\frac{d}{t^2}$$

Objectives

- **Investigate** Newton's second law of motion.
- **Manipulate** the CBL unit and the vernier motion detector to find the acceleration of an object down a ramp.
- **Observe** how mass affects an object's acceleration.

Materials

- laboratory cart
- 2 250-g masses
- 1000-g mass
- set of small masses
- pulley
- C-clamp
- heavy string (1.5 m)
- masking tape
- balance
- meterstick
- CBL unit
- vernier motion detector
- TI-92 calculator

Procedure

Part A

1. Set up a table and motion detector as shown in Figure A. Use masking tape to make distance marks at 0.5 m, 1 m, 1.5 m, and 2 m. Connect the CBL unit to the TI-92 calculator with the unit-to-unit link cable using the I/O ports located on the bottom edge of each unit. Press the cable ends in firmly. Connect the vernier motion detector to the SONIC port on the left side of the CBL unit. If necessary, obtain the L03ACCL program from your teacher and enter it in the TI-92. Place the motion detector on its side so that the gold foil lies directly on a distance mark and is 0.5 meter away from the path the cart will take.

6-1 Physics Lab

Figure A

2. Determine the combined mass of your laboratory cart and string. Record the mass in Table 1. Record the weight, $F = mg$, of the 1000-g mass in Table 2 as the accelerating force.

3. Assemble the cart, pulley, and string, as shown in Figure A. Tie a small loop in the end of the string hanging from the pulley.

4. Give the cart a small push toward the pulley. It should roll to a stop in a few centimeters. Attach a 10-g mass with a piece of masking tape to the end of the string that is hanging from the pulley. Give the cart another small push toward the pulley and observe its motion. If the cart moves at a constant velocity, then the weight of the 10-g mass is equal to the force of friction in the cart's wheels. If the cart rolled to a stop, then the total mass on the string must be increased. If the cart's velocity increased after it was released, then the total mass on the string must be decreased. Adjust the total mass until the cart moves at a constant velocity after you have given it a small push. Record in Table 1 the mass needed to equalize the friction of the cart. Leave the small masses attached to the string.

5. Begin the test for $d = 0.5$ meter. Carefully hang the 1000-g mass from the loop in the string. Adjust the string length so that the mass hangs just under the pulley. Hold the cart at the "start" line. Release the cart, allowing it to accelerate across the table top. Catch the cart before it collides with the pulley or plunges to the floor. After the graph has been made, on the TI-92, press 2nd [DRAW] 4 to get a vertical line on the screen. Use arrow keys to align this line with the *beginning* of the points that represent the cart. The displayed *x*-value gives you an estimation of the time it took the cart to pass the distance mark. Repeat this two more times. Find the average. Record your data in Table 2.

6. Repeat the procedure for distances of 1 meter, 1.5 meters, and 2 meters. Record your data in Table 2. Determine the acceleration of the cart for each trial. Record your results in Table 3.

Part B

7. Add a 250-g mass to the cart. Place the timer at the 1-meter mark and repeat the procedure. Record your results in Table 4.

8. Add another 250-g mass to the cart. Repeat step 7. Record your results in Table 4.

9. Determine the acceleration of the cart with each additional mass and record it in Table 4.

6-1 Physics Lab

Data and Observations

Table 1	
	Value
Mass of laboratory cart (kg)	
Mass needed to equalize friction of cart (kg)	

Table 2			
Trial	**Distance (m)**	**Accelerating Force (N)**	**Time (s)**
1	0.5 meter		
2	0.5 meter		
3	0.5 meter		
Average	0.5 meter		
1	1.0 meter		
2	1.0 meter		
3	1.0 meter		
Average	1.0 meter		
1	1.5 meter		
2	1.5 meter		
3	1.5 meter		
Average	1.5 meter		
1	2.0 meter		
2	2.0 meter		
3	2.0 meter		
Average	2.0 meter		

6-1 Physics Lab

Table 3			
Distance	Acceleration (m/s^2)	Total Mass (kg)	(m)(a) (N)
0.5 meter			
1.0 meter			
1.5 meter			
2.0 meter			

Table 4				
Trial	Distance	Acceleration (m/s^2)	Total Mass (kg)	(m)(a) (N)
empty cart	1.0 meter			
cart + 250 g	1.0 meter			
cart + 500 g	1.0 meter			

Analysis and Conclusions

1. Calculate the acceleration of the entire system using the values for distance and time from Table 2. Show your calculations for all trials. Record the average acceleration value for each distance in Table 3.

2. Are your acceleration values less than, equal to, or greater than g? Are these the values you would have predicted? Explain.

_____

_____

_____

_____

_____

6-1 Physics Lab

• • • • • • • • • • • • • • •

3. The total mass that was accelerated is equal to the mass of the cart, string, the small masses needed to equalize friction, and the 1000-g mass. Calculate the total mass of your system and record this value in Table 3.

4. Calculate the product of the total mass and the acceleration for each distance. Record these values in Table 3.

5. Compare your values of the product $(m)(a)$ to the accelerating force. Find the relative error using the accelerating force as the reference value.

_____

_____

6. Using your data, calculate the friction force.

7. Why was it important to neutralize the effect of the friction force acting on the system?

_____

_____

_____

6-1 Physics Lab

8. In Part B, how did the mass of the cart affect the acceleration?

_____

_____

Extension and Application

1. At a specific engine rpm, an automobile engine provides a constant force that is applied to the automobile. If the car is traveling on a horizontal surface, does the car accelerate when this force is applied? Explain.

_____

_____

_____

_____

_____

_____

_____

6-2

Design Your Own Physics Lab

Friction

An object on a slanted surface may or may not slide down, depending upon the angle of the slant. If the object remains at rest, the force of friction, F_f, between the object and the surface is sufficient to resist the weight, or force of gravity, pulling down on the object. One component of the weight is a force directed down the plane. This parallel component of weight is described by this equation.

$$F_\parallel = F_w \sin \theta$$

In the equation, the angle θ is the smallest angle, measured relative to the horizontal, at which sliding occurs. The perpendicular component of weight, $F_\perp$ is represented by this equation.

$$F_\perp = F_w \cos \theta$$

Figure A shows the friction force and the weight force, as well as the components of the weight force when a plane is inclined at angle θ.

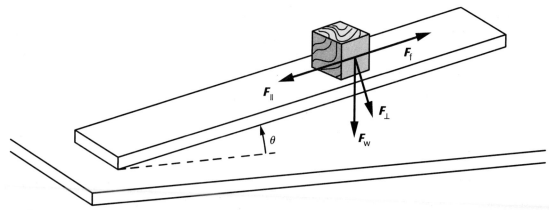

Figure A

An important value for this system of object and inclined plane is the coefficient of friction, μ. This value is represented by the following relationships.

$$\mu = \frac{F_f}{F_\perp} = \frac{F_\parallel}{F_\perp} = \frac{F_w \sin \theta}{F_w \cos \theta} = \tan \theta$$

The higher the value of the coefficient of friction is, the greater the friction force will be.

An object on an inclined plane is subject to two types of friction, with different coefficients of friction and different characteristic angles. Static friction exists when the object is at rest and must be overcome to start the object moving. To calculate the coefficient of static friction, you can use the equation above with θ equal to the angle at which motion begins.

If the object is already moving down the plane, sliding friction acts to resist the downward motion. To calculate the coefficient of sliding friction, use the equation above with θ equal to the angle at which the downward motion is at constant speed.

In this lab, you will design a procedure to measure the angle at which an object begins to slide down an inclined plane and, once already in motion, the angle at which it slides at constant speed. Finding

6-2 Physics Lab

those angles will give you the information you need to calculate the coefficients of static friction and sliding friction.

Problem

What are the coefficients of static friction and sliding friction for an object on an inclined plane?

Hypothesis

Formulate a hypothesis about the angles at which an object will begin to slide and slide at constant speed on an inclined plane, and about the relative magnitudes of the coefficients of static friction and sliding friction.

_____

_____

Plan the Experiment

1. Decide on a procedure that uses the suggested materials (or others of your choosing) to measure the force of sliding friction, which is the force needed to pull the object along the board at constant speed when the board is horizontal; the angle at which an object at rest begins to slide down an inclined plane; and the angle at which the same object, already in motion, slides at constant speed down an inclined plane.

2. Decide what data to collect and how to analyze it. You can record your data and calculated results in the table at the end of this lab. Make the appropriate columns.

3. Write your procedure on another sheet of paper or in your notebook.

4. **Check the Plan** Have your teacher approve your plan before you proceed with your experiment.

Analyze and Conclude

1. **Interpreting Data** Calculate the coefficient of sliding friction, using only data on weight and average force of sliding friction. Then calculate it using the angle of tilt and the equation involving the tangent of θ. Are the coefficients equal? Give reasons for any difference.

_____

_____

Objectives

- **Measure** the angle at which an object at rest begins to slide down an inclined plane.
- **Measure** the angle at which an object, already in motion, slides down an inclined plane at constant speed.
- **Calculate** the coefficients of static friction and sliding friction.

Possible Materials

- CBL unit
- link cable
- graphing calculator
- force sensor
- DIN adapter
- small rectangular object
- wooden board
- string
- masking tape
- protractor
- meterstick

6-2 Physics Lab

2. **Interpreting Data** Calculate the coefficient of static friction. How does this coefficient compare to the coefficient of sliding friction that you found?

_____

3. **Checking Your Hypothesis** Suppose you put a brick with its largest surface in contact with an inclined plane. You tilt the plane until the brick just begins to slide and measure the angle of the plane above the horizontal. Then you turn the brick on one of its narrow edges, tilt the plane, and measure the tilt angle. Will these measured angles be different? Explain your answer in terms of the equation for the force of friction.

_____

_____

_____

_____

_____

_____

_____

4. **Checking Your Hypothesis** You put a brick on an inclined plane and tilt the plane until the brick just begins to slide. You measure the angle of the plane with respect to the horizontal. You wrap the brick in wax paper and put it on the plane, tilt the plane, and measure the tilt angle. Will these measured angles be different? Give a reason for your answer.

_____

_____

_____

_____

_____

_____

5. **Inferring** Based on your answers to questions 3 and 4, describe the factors that influence the force of friction.

_____

_____

_____

_____

6-2 Physics Lab

Apply

1. While looking for a set of new tires for your car, you find an ad that offers two brands of tires at the same price. Brand X has a coefficient of friction of 0.90 on dry pavement and of 0.15 on wet pavement. Brand Y has a coefficient of friction of 0.88 on dry pavement and of 0.45 on wet pavement. If you live in a rainy area, which tire would give you better service? Give a reason for your answer.

_____

_____

_____

_____

Data and Observations

Data Table

6-3

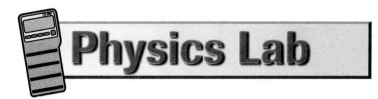

Physics Lab

Pushes, Pulls, and Vectors

Objectives

- **Measure** the force applied in moving an object.
- **Graph** the force and the direction in which it was applied as a vector.
- **Analyze** the vector's horizontal component.

Pushes and pulls can be described by vectors. Think about trying to pull an object across the ground with the help of a rope. You might pull with the rope parallel to the ground. You might hold the rope up towards your chest while pulling. You might even turn around, put the rope over your shoulder, and tug on it that way. In each case, you are applying some horizontal force on the rope. By analyzing the vector of force you used, you can find out how much horizontal force you applied.

Vernier student force sensor

String

Book 45°

TI-82   CBL

Figure A

Materials

- meterstick
- protractor
- masking tape
- book
- string
- CBL unit
- vernier student force sensor with CBL DIN adapter
- TI-82 calculator
- TI-92 calculator

Procedure

Part A

1. Connect the CBL unit to the TI-82 calculator with the unit-to-unit link cable using the I/O ports located on the bottom edge of each unit. Press the cable ends in firmly. Connect the vernier student force sensor to Channel 1 (CH1) on the top edge of the CBL unit. If necessary, calibrate the force probe. (See the vernier student force sensor guide for details.) If necessary, obtain the L07FORCE program from your teacher and enter it into the TI-82.

2. Tie one end of the string around the book so that it is easy to pull without tipping over. You may have to attach a separate string to two or more points on the book in order to stabilize your pull and tie the main string to those separate strings. Attach the other end of the string to the vernier student force sensor.

3. Use masking tape to put two marks on a flat area—a 0 mark and a mark 0.5 m from the 0. Place the book so that the edge attached to the string is at the 0 mark. Determine an angle of 45° above the horizontal. Hold the string at that angle.

6-3 Physics Lab

• • • • • • • • • • • • • •

4. Turn on the CBL and the TI-82. Start the L07FORCE program on the TI-82. After starting the force graph, begin pulling steadily on the string in the direction of the determined angle. Be sure the force meter forms a right angle with the string. Use a consistent force to pull the book 0.5 m in 5 seconds. Read the constant force on the TI-82. (You will be able to tell when you have reached the constant force when the line on the student force sensor becomes horizontal. You may want to change Ymax and Ymin on the TI-82 to adjust the viewing window.)

5. On the TI-82, press 2nd [DRAW] 3 to get a horizontal line on the screen. Use the arrow keys to align this line with the constant force. The displayed *y* value gives you an estimation of the average constant force applied during this time. Record this value in Table 1.

6. Repeat steps 3–5 for angles of 30°, 15°, and 0°.

Part B

1. Use the TI-92. Draw the vector that represents the force you applied at an angle of 45°. Draw the downward vertical component vector on the same graph, putting the tail of this vector at the head of the initial vector. Compute the horizontal force used. Record this force in Table 1.

2. Repeat step 1 for the data at the angles of 30°, 15°, and 0°. Write your results in the table.

Data and Observations

Table 1		
	Amount of force applied (*y*-value)	**Amount of horizontal force applied**
45°		
30°		
15°		
0°		

Analysis and Conclusions

1. Would the horizontal force you apply while pulling at a 30° angle be identical to the horizontal force you apply at a −30° angle (pulling down from the horizontal)?

_____

_____

_____

_____

_____

_____

6-3 Physics Lab

2. Review the total force applied and the horizontal force applied for each angle. What pattern do you see?

_____

_____

_____

_____

_____

_____

3. Review your data. How difficult do you think it would be to pull the book if the force were applied at 60°? Explain how you arrived at your conclusion.

_____

_____

_____

_____

_____

_____

_____

_____

4. Would a force applied at 90° produce a horizontal force that is one-half the horizontal force applied at 45°? Why or why not?

_____

_____

_____

_____

_____

_____

_____

6-3 Physics Lab

Extension and Application

1. A participant in a tug-of-war generally pulls at an angle of 0°. Why might it be easier for a person to move some objects, for example a small wagon, using a 45° pull rather than a 0° pull? You may use the CBL, TI-82, and vernier student force sensor to find out.

_____

_____

_____

_____

7-1

Physics Lab

Projectile Motion

Objectives

- **Investigate** the path of a projectile.
- **Verify** equations of projectile motion.
- **Graph** the relationships.

An object that is launched into the air and then comes under the influence of gravity moves in two dimensions and is called a projectile. If the frictional force due to air resistance is disregarded, the horizontal component of velocity will remain constant during the projectile's entire flight. The equation for the horizontal displacement of an object is the following.

$$x = v_x t$$

In the equation, x is the horizontal displacement, v_x is the initial horizontal velocity, and t is the elapsed time. The vertical component of velocity is the same as the motion of an object in free fall. The force due to gravity accelerates the projectile downward at the rate of 9.80 m/s². The vertical displacement of an object falling with constant acceleration, g, is described by the following relationship.

$$y = v_y t + \frac{1}{2} g t^2$$

In this equation, y is the vertical displacement, v_y is the initial vertical velocity, t is the elapsed time, and g is the acceleration due to gravity.

In Figure A, a steel ball is projected horizontally from the bottom of the raised ruler and rolls down an inclined plane across the paper. The acceleration of the steel ball is the component of the acceleration due to gravity that acts parallel to the direction of the inclined plane. The projectile's horizontal velocity remains nearly constant because the frictional effects of the steel ball on the smooth paper are negligible.

Figure A

The path of the steel ball will be marked on the lined paper in a pattern similar to the one shown by the dashed line.

Procedure

1. Set up the apparatus as shown in Figure A. Lay the lined paper on the board or tray. Position the paper so that the groove in the raised ruler is perpendicular to the vertical lines and tape the corners to the board. Adjust the position and height of the ruler so that the steel ball will begin its path at the upper left-hand corner and travel most of the way across the paper. Secure the ruler and block with tape when the ball rolls the proper distance.

7-1 Physics Lab

Name _____

Materials

- steel ball
- 1 sheet lined paper
- masking tape
- board or tray
- grooved metric ruler
- 1 sheet carbon paper
- textbook support
- block support
- TI-92 calculator
- stopwatch

2. Practice placing the steel ball at different heights on the raised ruler until you have found the best location for releasing it and obtaining the desired result. When you have found the best location, make a few practice trials using a timer. Start the trial at time $t = 0$, and record the time for the entire trial.

3. Place the carbon paper over the lined sheet with the carbon side down and tape the top corners to hold it in place. As the steel ball moves across the carbon paper, it will trace its path on the lined paper.

4. Start the timer and simultaneously roll the steel ball down the grooved ruler and across the carbon paper. Stop the timer as soon as the ball reaches the bottom of the paper. Lift the carbon paper and check to see that the carbon paper left a trace of the ball's path. Remove the carbon paper and the lined paper from the board. With a pencil, retrace the ball's path if it is difficult to see. The path should resemble the path in Figure B.

5. Because the horizontal velocity of the ball was constant, it took the ball the same time to travel each of the horizontal distances between adjacent vertical lines on the paper. Therefore, each width of a section on the paper represents successive intervals of equal time.

   Total time ÷ the number of time intervals = the time of each interval.

Figure B

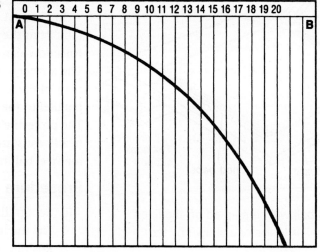

6. Measure in centimeters the vertical distance along each line from the horizontal line **AB** to the path of the ball. Record each of these distances in Table 1.

7. Determine the distance the ball traveled down the incline during each time interval. Using the values for vertical distance, subtract the length on the previous vertical line from the length on the current vertical line to determine the distance the ball traveled in the current interval. The distance the ball traveled during each interval represents its average vertical velocity. Record each value for average vertical velocity in Table 1.

7-1 Physics Lab

Data and Observations

Table 1			
Interval number	Time of interval (s)	Vertical distance (m)	Average vertical velocity (cm/s)
0			
1			
2			
3			
4			
5			
6			
7			
8			
9			
10			
11			
12			
13			
14			
15			
16			
17			
18			
19			
20			

Physics: Principles and Problems

7-1 Physics Lab

Analysis and Conclusions

1. Plot a graph of vertical distance versus time, with vertical distance on the y-axis and time on the x-axis.

2. What does this graph show about projectile motion?

   _____

   _____

3. What other data do you have that show the path of the steel ball in a form consistent with the shape of your graphed line from question 1? Explain.

   _____

   _____

4. Using your data and the equation for the horizontal displacement of a projectile, $x = v_x t$, find v_x. Using the TI-92 calculator, graph the value of v_x over time. Describe the graph.

   _____

   _____

5. Plot a graph of vertical velocity versus time, with vertical velocity on the y-axis and time on the x-axis.

6. What does this graph indicate about projectile motion?

   _____

   _____

7. Using your data and the equation for vertical displacement of a falling object, $y = v_y t + \frac{1}{2} gt^2$, find v_y. Using the TI-92 calculator, graph the value of v_y over time. Describe the graph.

   _____

   _____

Extension and Application

1. A man raises his gun and aims it at a tin can target on a shelf in a shooting range. If the bore of the gun is pointed straight at the can, under what conditions will the bullet hit the tin can?

   _____

   _____

   _____

   _____

7-2

Design Your Own Physics Lab

Range of a Projectile

Objectives

- **Measure** the time a projectile takes to travel a certain distance.
- **Calculate** horizontal velocity of a projectile.
- **Predict** the horizontal distance a projectile will travel.
- **Verify** that horizontal and vertical motion components are independent of each other.

Consider a leaping cat, a rocket launched from Earth, or a golf ball hit from a tee. Such objects, called projectiles, move in two dimensions. The horizontal and vertical components of the motion are independent of each other and determine the location and time of fall of projectiles. All three of the projectiles just mentioned have an initial velocity at an angle above the horizontal, which complicates the situation somewhat. In this lab, to simplify analysis of the motion of the projectile (a steel ball), the ball will be launched horizontally. Therefore, its initial vertical velocity, v_y, will be zero.

If you know the horizontal velocity, v_x, of a steel ball launched horizontally from a table and the ball's initial height, y, above the floor, you can use the equations of projectile motion to predict where the steel ball will land. The horizontal displacement of an object with horizontal velocity v_x at time t is represented by the following equation.

$$x = v_x t$$

The equation describing vertical displacement is that of a body falling with constant acceleration.

$$y = v_y t + \frac{1}{2} gt^2$$

You can see that if $v_y = 0$, then $y = gt^2/2$, where g is the acceleration due to gravity, 9.80 m/s², and y is the vertical height the projectile falls.

To compute the steel ball's fall time from the equation for vertical motion, solve for t.

$$t = \sqrt{\frac{2y}{g}}$$

Then you can compute the range of the projectile, x.

Possible Materials

-
- 2 pieces of U-shaped channel
- steel ball
- masking tape
- ring-stand support clamp
- meterstick
- string
- washer
- stopwatch
- paper cup

Problem

Where will a horizontally projected object land?

Hypothesis

Formulate a hypothesis as to whether you can predict where a steel ball launched from a horizontal ramp will land.

_____

_____

7-2 Physics Lab

Plan the Experiment

1. Decide on a procedure that uses the suggested materials (or others of your choosing) to obtain measurements that will enable you to calculate the horizontal distance a steel ball will travel before hitting the floor.

2. Decide what kind of data to collect and how to analyze it to calculate and test a predicted value for horizontal range. You can record your data and calculated results in the table below. Label the columns appropriately.

3. Write your procedure on a separate sheet of paper or in your notebook. On page 51, draw the set-up you plan to use.

4. **Check the Plan** Have your teacher approve your plan before you proceed with your experiment. CAUTION: *Do not leave dropped steel balls on the floor, where someone could slip on them.*

Data and Observations

Data Table				

7-2 Physics Lab

Setup

Analyze and Conclude

1. **Analyzing Results** Did the ball land where you predicted? If not, explain why it didn't.

   _____

   _____

   _____

   _____

2. **Checking Your Hypothesis** You jump from a diving board with a horizontal velocity of 3.75 m/s. If the diving platform is 6.0 m above the water, at what horizontal distance from the platform will you hit the water?

7-2 Physics Lab

• • • • • • • • • • • • • • •

3. **Inferring** If a light but fairly large sponge ball or a Ping-Pong ball were used instead of the steel ball, would you expect the same result? Give a reason for your answer.

_____

_____

_____

_____

_____

Apply

1. The horizontal range, x, of a projectile launched at an angle θ_i above the horizontal at an initial velocity v_i can be determined using this equation.

$$x = \frac{v_i^2}{g} (\sin 2\theta_i)$$

Use the equation to find the horizontal range of a baseball that leaves the bat at an angle of $63°$ with the horizontal and an initial velocity of 160 km/h. Disregard air resistance. *Hint:* Simplify your calculations by using the following relationship, which applies for any angle θ.

$$\sin 2\theta = 2 \sin \theta \cos \theta$$

7-3 Physics Lab

Torques

Objectives

- **Measure** forces that produce torque.
- **Calculate** clockwise and counterclockwise torque on a rotating body.
- **Determine** the relationship between torque and lever-arm distance.

There is a good reason for attaching a doorknob as far as possible from the hinges of the door. When you open a door, you must apply a force. Where you apply that force and in what direction you push or pull determine how easily the door will rotate open. Forces not only produce motion in a straight line, they also produce rotation in a rigid body that has an axis. The turning movement caused by one or more forces acting on a body that is free to rotate about an axis is known as torque, τ, and is defined as the product of the vector quantity force, F, and lever arm, r. The lever arm is the perpendicular distance from the axis of rotation to a line drawn in the direction of the force. Thus, the farther this line is from the axis, the more effective is the force that causes rotation. Work, W, is the product of force and the distance through which the force acts, so the work done in rotating an object through an angle $\Delta\theta$ is $W = (F)(r\Delta\theta)$. But $Fr = \tau$; therefore, $W = \tau\Delta\theta$.

Figure A

In this experiment, two parallel forces, the two spring scale balances, will balance the downward force of the hanging mass, as shown in Figure A. The two parallel forces will tend to cause rotation because each is exerted over a specific distance, the lever arm, from the third force, the weight of the hanging mass. The weight force is at the pivot, point B in the figure. This fixed point B is the axis of rotation.

The force at point A acts over a distance AB and produces a clockwise torque (a negative value). The force at point C acts over a distance BC and

7-3 Physics Lab

produces a counterclockwise torque (a positive value). The sum of these two torques may cause the meterstick to rotate in either a clockwise or counterclockwise direction about the axis. If the sum of all the clockwise torques about the axis and the sum of all the counterclockwise torques about the same axis equal zero, the meterstick will not rotate. In this experiment, when the two parallel forces are balanced by the third force, the sum of all torques will equal zero, and the system will be in rotational equilibrium.

Materials

meterstick

2 spring scales

3 meterstick clamps

500-g hooked mass

masking tape

Procedure

1. Set up the apparatus as shown in Figure A, without the hanging mass. Hang the two spring balances from supports on the laboratory table or tape them with masking tape so they hang over the edge of the table. Be sure that the spring scale mechanisms can move freely.

2. For the first trial, center one clamp at the 5-cm mark on the meterstick (point A) and center the other clamp at the 95-cm mark (point C), as shown in Figure A.

3. Observe the force readings on the spring scales. Record these values in Table 1 as the original readings.

4. Hang a 500-g mass (4.9 N) from the clamp at point B, at the center of the meterstick. Observe the scale readings at points A and C when the apparatus is in equilibrium. Record these values in Table 1 as the final readings.

5. The real reading of each spring scale is the difference between the final reading and the original reading. The real reading is the force due to the hanging mass at point B. Calculate the real reading for each scale and record these values in Table 1.

6. Measure and record the distances AB and BC in Table 2.

7. The clockwise torque is equal to the product of the real reading for spring scale a and the distance AB. The counterclockwise torque is equal to the product of the real reading for spring scale c and the distance BC. Calculate the clockwise and counterclockwise torques and record these values in Table 2.

8. Repeat steps 2–7 for trials 2 and 3, moving the clamp at point A to two different positions on the meterstick.

Copyright © by Glencoe/McGraw-Hill

7-3 Physics Lab

Data and Observations

Table 1					
	Balance a			Balance c		
Trial	Original reading (N)	Final reading (N)	Real reading (N)	Original reading (N)	Final reading (N)	Real reading (N)
1						
2						
3						

Table 2			
Trial	Distance AB (m)	Distance BC (m)	Clockwise torque (N·m)	Counterclockwise torque (N·m)
1				
2				
3				

Analysis and Conclusions

1. Because the system in each trial was in equilibrium, what conditions had been met?

   _____

   _____

   _____

2. What is the relationship between the magnitudes of the forces exerted (real reading) and the lever-arm distances over which the forces were exerted?

   _____

   _____

7-3 Physics Lab

3. Compare the absolute values of the clockwise and counterclockwise torques for each trial by finding the relative percentage difference between the two values.

$$\% \text{ difference} = \frac{|\text{counterclockwise torque}| - |\text{clockwise torque}|}{|\text{counterclockwise torque}|} \times 100\%$$

4. What is the relationship between the counterclockwise torque and the clockwise torque when the system is in equilibrium?

   _____

   _____

5. When the systems in your trials were in equilibrium, how much work was done? Give a reason for your answer.

   _____

   _____

   _____

Extension and Application

1. The technique for measuring forces in this lab can be used to determine the forces exerted by the piers of a bridge. In this case, the forces are reversed. The two pier supports (replacing the two spring scales) exert an upward force, and a person standing on the bridge (replacing the hanging mass) exerts a downward force on the bridge. Calculate the force exerted by each pier (F1 and F2) of a 5.0-m footbridge having a mass of 100.0 kg, evenly distributed, when a 55-kg person stands 2 m from one end of the bridge.

8-1

Physics Lab

Kepler's Laws

The motion of the planets has intrigued astronomers since they first gazed at the stars, moon, and planets filling the evening sky. But the models using eccentrics and equants (combinations of circular motions) did not accurately account for planetary movements. Johannes Kepler adopted the Copernican theory that Earth revolves around the sun (the heliocentric, or sun-centered, view) and closely examined Tycho Brahe's meticulously recorded observations on Mars's orbit. With these data, he concluded that Mars's orbit is not circular. With elliptical orbits as a model, all the discrepancies of planetary motion disappeared. From his studies, Kepler derived three laws that apply to every satellite or planet orbiting another massive body.

1. The paths of the planets are ellipses, with the center of the sun at one focus.

2. An imaginary line from the sun to a planet sweeps out equal areas in equal time intervals, as shown in Figure A.

3. The ratio of the squares of the periods of any two planets revolving about the sun is equal to the ratio of the cubes of their respective average distances from the sun. Mathematically, this relationship can be expressed as

$$\frac{T_a^2}{T_b^2} = \frac{r_a^3}{r_b^3}.$$

In this lab, you will use heliocentric data to plot the positions of Mercury. Then you will draw Mercury's orbit. The distance from the sun, the radius vector, is compared to Earth's average distance from the sun, which is defined as 1 astronomical unit or 1 AU. The angle, or longitude, between the planet and a reference point in space is measured from the zero degree point, or vernal equinox.

Objectives

- **Plot** a planet's orbit.
- **Apply** Kepler's laws of planetary motion.
- **Calculate** a planet's average distance from the sun.
- **Compare** the orbits of two planets.

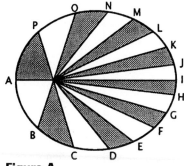

Figure A
Kepler's law of areas

Materials

- polar graph paper
- sharp pencil
- metric ruler

Procedure

1. Orient your polar graph paper so that the zero degree point is on your right. The sun is at the center of the paper. Label the sun without covering the center mark. Move about the center in a counterclockwise direction as you measure and mark the longitude.

2. Select an appropriate scale to represent the values for the radius vectors of Mercury's positions. Since Mercury is closer to the sun than Earth is, the radius vector always is less than 1 AU. In this step, then, each concentric circle could represent 0.1 AU.

3. Table 1 provides the heliocentric positions of Mercury over a period of several months. Start with the data for October 1 and locate the longitude on the polar graph paper. Measure along the longitude

8-1 Physics Lab

• • • • • • • • • • • • • •

line an appropriate distance, in your scale, for the radius vector for this date. Make a small dot at this point to represent Mercury's distance from the sun. Write the date next to this point.

4. Repeat step 3, plotting all longitudes and radius vectors.

5. After plotting all the data, carefully connect the points to show the orbit of Mercury.

Data and Observations

Table 1					
Some Heliocentric Positions for Mercury for 0^h Dynamical Time*					
Date	Radius vector (AU)	Longitude (degrees)	Date	Radius vector (AU)	Longitude (degrees)
Oct. 1, 1990	0.319	114	Nov.16	0.458	280
3	0.327	126	18	0.452	285
5	0.336	137	20	0.447	291
7	0.347	147	22	0.440	297
9	0.358	157	24	0.432	304
11	0.369	166	26	0.423	310
13	0.381	175	28	0.413	317
15	0.392	183	30	0.403	325
17	0.403	191	Dec. 2	0.392	332
19	0.413	198	4	0.380	340
21	0.423	205	6	0.369	349
23	0.432	211	8	0.357	358
25	0.440	217	10	0.346	8
27	0.447	223	12	0.335	18
29	0.453	229	14	0.326	29
31	0.458	235	16	0.318	41
Nov. 2	0.462	241	18	0.312	53
4	0.465	246	20	0.309	65
6	0.466	251	22	0.307	78
8	0.467	257	24	0.309	90
10	0.466	262	26	0.312	102
12	0.464	268	28	0.319	114
14	0.462	273	30	0.327	126

* Adapted from *The Astronomical Almanac for the Year 1990,* U.S. Government Printing Office, Washington, D.C., 20402, p. E9.

8-1 Physics Lab

Analysis and Conclusions

1. Does your graph of Mercury's orbit support Kepler's law of orbits?

   _____

   _____

2. Draw a line from the sun to Mercury's position on December 20. Draw a line from the sun to Mercury's position on December 30. The two lines and Mercury's orbit describe an area swept by an imaginary line between Mercury and the sun during the 10-day interval. Lightly shade this area. Over a small portion of an ellipse, the area can be approximated by assuming the ellipse is similar to a circle. The equation that describes this value is

   $$\text{area} = (\theta/360°)\pi r^2,$$

   where r is the average radius for the orbit.

   Determine θ by finding the difference in degrees between December 20 and December 30. To approximate the area, measure the radius at a point midway in the orbit between the two dates. Calculate the area in AUs for this 10-day period.

3. Select two additional 10-day periods at points distant from the interval in question 2 and shade these areas. Approximate the area in AUs for each period.

4. Find the average area for the three periods of time from questions 2 and 3. Calculate the relative error between each area and the average. Does Kepler's law of areas apply to your graph?

8-1 Physics Lab

5. Calculate the average radius for Mercury's orbit by averaging all the radius vectors that occur along the major axis. The major axis is shown in Figure B. Recall that the sun is at one focus; the other focus is a point that is the same distance from the center of the ellipse as the sun but in the opposite direction.

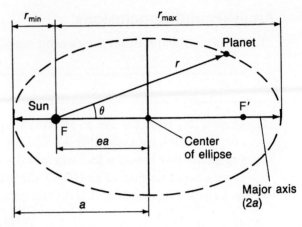

Figure B

The major axis passes through the two foci (F and F') and the center of the ellipse. The value *ea* determines the location of the foci; *e* is the eccentricity of the orbit. If $e = 0$, the orbit would be a circle, and the foci would merge at one central point.

From Table 1, find the longest radius vector. Then, align a metric ruler so that it passes through the point on the orbit that represents the longest radius vector and through the center of the sun to a point opposite on the orbit. Find the shortest radius vector by reading the longitude at this opposite point and consulting Table 1 for the corresponding radius vector. Average these two radius vector values. Using the values for Earth's average radius (1.0 AU), Earth's period (365.25 days), and your calculated average radius of Mercury's orbit, apply Kepler's third law to find the period of Mercury.

6. Refer again to the graph of Mercury's orbit. Count the number of days it took Mercury to orbit the sun; recall that this orbital time is the period of Mercury. Is this value different from the value you calculated in question 5? Calculate the relative difference between these values. Are the results from your graph consistent with Kepler's law of periods?

8-1 Physics Lab

Extension and Application

1. There has been some discussion about a hypothetical planet X that is on the opposite side of the sun from Earth and that has an average radius of 1.0 AU. If this planet exists, what is its period? Show your calculations.

2. Using the data in Table 2, plot the radius vectors and corresponding longitudes for Mars. Does the orbit you drew support Kepler's law of ellipses? Select three different areas and find the area per day for each of these. Does Kepler's law of areas apply to your model of Mars?

Table 2					
Some Heliocentric Positions for Mars for 0^h Dynamical Time*					
Date	Radius vector (AU)	Longitude (degrees)	Date	Radius vector (AU)	Longitude (degrees)
Jan. 1, 1990	1.548	231	July 12	1.382	343
17	1.527	239	28	1.387	353
Feb. 2	1.507	247	Aug. 13	1.395	3
18	1.486	256	29	1.406	13
Mar. 6	1.466	265	Sept. 14	1.420	23
22	1.446	274	30	1.436	32
Apr. 7	1.429	283	Oct. 16	1.455	42
23	1.413	293	Nov. 1	1.474	51
May 9	1.401	303	17	1.495	60
25	1.391	313	Dec. 3	1.516	68
June 10	1.384	323	19	1.537	76
26	1.381	333	Jan. 4, 1991	1.557	84

* Adapted from *The Astronomical Almanac for the Year 1990*, U.S. Government Printing Office, Washington, D.C., 20402, p. E12.

8-1 Physics Lab

9-1

Physics Lab

Conservation of Momentum

The law of conservation of momentum states: The momentum of any closed, isolated system does not change. This law is true regardless of the number of objects or directions of the objects before and after they collide. In a collision or interaction, the momentum before the collision equals the momentum after the collision.

The change in an object's momentum is equal to the product of the force acting on the object and the interval of time the force acts. In this investigation, two carts will be pushing away from each other. Thus, from Newton's third law of motion, each cart will give the other an equal, but opposite, impulse. The change of momentum from the first cart will equal the change of momentum of the second cart. The two carts in this investigation will have unequal masses. The two carts will be placed together compressing a spring in one of the carts. The carts will move apart when the spring is released.

Objectives

- **Measure** the masses and velocity of two carts.
- **Calculate** the momentum of the two carts.
- **Apply** the conservation of momentum to a system.
- **Demonstrate** conservation of momentum for an interaction.

Materials

- safety goggles
- 2 collision carts, one with a spring-mechanism
- balance
- meterstick
- CBL unit
- ultrasonic motion detector
- link cable
- graphing calculator
- set of masses
- masking tape
- index cards, 5 × 7 inches
- ring stand
- clamp

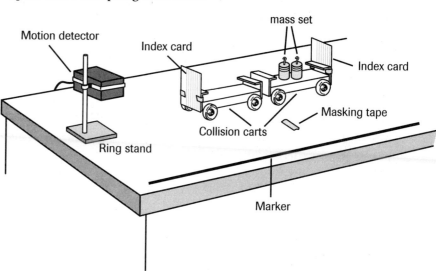

Procedure

1. Select a counter space having 2–3 meters of clear space. Place a piece of masking tape in the middle of this space.

2. Tape an index card onto each of the collision carts. This will make the carts more easily seen by the ultrasonic motion detector. Mass each cart and record this value in Table 1. Using your mass set, add enough masses to cart 2 so its mass is twice the mass of cart 1. It is desirable to have the mass of cart 2 within 5% of twice its original mass. Record the mass added to cart 2 in Table 1. Secure the masses to the cart with a small piece of masking tape. Compress the spring mechanism and place the carts centered over the tape marker.

9-1 Physics Lab

3. Mount the ultrasonic motion detector on a ring-stand clamp, using tape if necessary. Adjust the height of the sensor so it is at the same height as the middle of the index card. Place the motion detector about 1.5 meters away from the masking-tape marker.

4. Connect the CBL unit to the calculator, using the unit-to-unit link cable and the link ports on the calculator and CBL unit. Connect the ultrasonic motion detector to the SONIC port of the CBL unit.

5. Turn on the CBL unit and graphing calculator. If not already loaded into your calculator, load the program PHYSICS and its subprograms into your calculator from another calculator or download them from a computer. Start the program PHYSICS on your graphing calculator.

6. Select the option SET UP PROBES from the MAIN MENU. Enter 1 for the number of probes. Then from the SELECT PROBE menu, choose MOTION from the list.

7. Select the MONITOR INPUT option from the COLLECT DATA menu. This will allow you to check and see if the motion detector is working properly by displaying the distance between the motion detector and the vehicle on the CBL unit. Press the "+" to quit the sampling test.

8. You are ready to begin the experiment. Select the COLLECT DATA option. In DATA COLLECTION, select TIME GRAPH. The calculator will prompt you to ENTER TIME BETWEEN SAMPLES IN SECONDS. Enter 0.03 for the time between samples. Enter 99 for the number of samples. (A TI-82 can collect only 99, a TI-85 can collect only 55, while a TI-83 can collect 120.) Check the values you entered and press ENTER. If the values are correct, select USE TIME SETUP to continue. If you have made a mistake entering the values, then select MODIFY SETUP and reenter the values before continuing.

9. On the TIME GRAPH menu, select NON-LIVE DISPLAY.

10. Press ENTER. The READY EQUIPMENT command should appear. While preventing the carts from moving, one student should press ENTER on the graphing calculator. When the motion detector begins to click, depress the spring-release mechanism to release the carts. Stop the carts before one strikes the motion detector or falls on the floor.

11. When the motion detector has stopped clicking and the CBL unit displays DONE, press ENTER on the graphing calculator. From the SELECT GRAPH menu, select DISTANCE to plot a graph of the distance in meters against the time in seconds.

12. The displacement graph will initially show a horizontal line until the cart begins to move. The beginning of the graph will show where the cart accelerated as the spring was released. Use the right arrow key to move along the graph to the middle section where the cart was moving at a relatively constant speed. Select a point near the beginning of this middle section and one near the end. Record the distance (y-value) and the time (x-value) into your data table for each point. Press ENTER to return to the SELECT GRAPH menu.

13. Rotate the two carts so the second cart is facing the motion detector. Repeat the experiment collecting data for the second cart.

14. Repeat the experiment for a second trial.

Copyright © by Glencoe/McGraw-Hill

9-1 Physics Lab

Data and Observations

Table 1		
	Mass	**Mass Added (kg)**
Cart 1		--
Cart 2		

Table 2							
	Begin t	**End t**	**Δt**	**Begin d**	**End d**	**Δd**	**Average velocity**
Cart 1-trial 1							
Cart 1-trial 2							
Cart 2-trial 1							
Cart 2-trial 2							

Analysis and Conclusions

1. For each trial, find the average velocity of each cart and enter these in Table 2.

2. For each trial, calculate the momentum of each cart.

_____
_____
_____
_____

Name _____

3. Calculate the total momentum of the two carts before the spring mechanism was released.

4. Was the velocity of the carts equal before and after the spring-mechanism release?

_____

_____

5. Calculate the final total momentum. Compare the initial and final momentum of the carts.

_____

_____

_____

Extension and Application

1. While playing baseball with your friends, your hands begin to sting after you catch several fast balls. What method of catching the ball might prevent this stinging sensation?

_____

_____

_____

2. If an incident ball A hit two target balls, B and C, at an angle, predict what would happen to the total momentum of the system after the collision. Give a reason for your answer and write an equation that proves your answer.

_____

_____

_____

_____

3. When an incident ball, A, collides at an angle with a target ball, B, of equal mass that is initially at rest, the two balls always move off at right angles to each other after the collision. Use a familiar equation for a right triangle to show that this statement is true. *Hint:* Since the collision is elastic, kinetic energy is conserved and

$$\frac{1}{2}\, mv_{A1}^2 = \frac{1}{2}\, mv_{A2}^2 + \frac{1}{2}\, mv_{B2}^2.$$

9-2

Physics Lab

Angular Momentum

Newton's second law applies to rotational motion as well as to linear motion. The rotational analog of $F = ma$ is $\tau = I\alpha$, where τ is the torque, the force applied to an object rotating about a fixed axis; I is the rotational inertia of the object; and α is the angular acceleration. The rotational inertia of an object is its resistance to changes in its angular velocity. Rotational inertia involves the mass of a rotating object and the distribution of its mass, or its shape.

Like other linear values, linear momentum has an angular equivalent. In linear motion, momentum is equal to mv. Substituting the rotational equivalents of m and v, the angular momentum, L, is

$$L = I\omega,$$

where ω is the angular velocity. If we look at the change in angular momentum over a period of time, the equation can be written

$$\frac{\Delta L}{\Delta t} = I\left(\frac{\Delta\omega}{\Delta t}\right).$$

Since $\alpha = \Delta\omega/\Delta t$, $\tau = I\alpha = I(\Delta\omega/\Delta t)$. Substituting $\Delta L/\Delta t$ for $I(\Delta\omega/\Delta t)$ yields

$$\tau = \frac{\Delta L}{\Delta t} = \text{change in angular momentum/time interval.}$$

Note that if the torque is zero, then ΔL must also be zero. If there is no change in the angular momentum over a period of time, it must be conserved. If there are no net external torques acting on an object, the angular momentum, $I\omega$, is a constant. While I and ω can change, their product must remain constant. This relationship can be written

$$I_1\omega_1 = I_2\omega_2.$$

For example, if an object is rotating at a given velocity about a fixed axis, its rotational inertia must increase if its angular velocity decreases and vice versa. Angular momentum is a vector quantity, and its direction is along the axis of rotation perpendicular to the plane formed by ω and r, where r is the object's radius of rotation.

Procedure

Use caution while performing the following activities. If you become dizzy or nauseated, you may lose your balance and fall off the stool or platform.

1. Sit on a rotating stool, or stand on a rotating platform, and hold a large mass in each hand. Extend your arms and have your lab partner give you a gentle spin. Observe what happens as you pull your

Objectives

- **Observe** the torque produced by rotational inertia.
- **Apply** the law of conservation of momentum.
- **Predict** the effects of rotational inertia.

Materials

- gyroscopic bicycle wheel
- rotating stool or rotating platform
- 2 masses, 3 kg each
- safety goggles

9-2 Physics Lab

arms in. Record your observations in item 1 of Data and Observations.

2. Hold one side of the bicycle wheel shaft. Slowly tilt the wheel upward (the wheel should not be rotating). What happens? Record your observations in item 2 of Data and Observations. Hold the gyroscopic bicycle wheel by the shaft with both hands. Have your lab partner give the wheel a strong spin. Let go of one side of the shaft, as shown in Figure A. Slowly tilt the shaft upward.

   What happens? Record your observations in item 3 of Data and Observations. Quickly tilt the shaft upward or downward. What happens? Record your observations in item 3.

(1)

(2)

Figure A

(1) The angular momentum vector extends outward along the shaft from the axle of a spinning gyroscopic bicycle wheel. (2) When the shaft is tilted in time Δt, the angular momentum changes by ΔL.

3. While sitting on the rotating stool or standing on the rotating platform, hold the gyroscopic bicycle wheel shaft with both hands, as shown in Figure B. Have your lab partner spin the wheel. Slowly rotate the wheel to the right by raising your left hand and lowering your right hand. Observe what happens. Record your observations in item 4.

4. Change positions with your lab partner so that each student has a set of observations.

Figure B

Data and Observations

1. Observations of student holding masses on spin-ning stool or platform:

_____

_____

_____

_____

2. Observations of effects of tilting the shaft of a stationary wheel upward:

_____

_____

_____

_____

9-2 Physics Lab

3. Observations of effects of tilting the shaft of a rotating wheel upward:

_____

_____

_____

_____

4. Observations of effects of rotating the spinning wheel:

_____

_____

_____

_____

Analysis and Conclusions

1. According to the observations you recorded in item 1, is your angular momentum conserved? Explain, using the law of conservation of momentum.

_____

_____

_____

_____

2. Under what conditions could the angular momentum of the closed system of isolated student, stool, and masses change? Describe one such condition.

_____

_____

_____

_____

3. Compare the effects of tilting the shaft of a stationary wheel and of a rotating wheel. Using your observations in items 2 and 3 and the relationship between torque and angular momentum, explain the result of your attempt to change the angular momentum of the rotating wheel.

_____

_____

_____

_____

9-2 Physics Lab

• • • • • • • • • • • • • •

4. Using your observations in item 4, explain what happened in terms of the law of conservation of momentum.

_____

_____

_____

Extension and Application

1. If a uniformly filled cylinder, such as a solid wood cylinder, and a hollow cylinder or hoop, such as a can with clay pressed on the inside edge, roll down an inclined plane, would you expect them to reach the bottom at the same time? Try it. Explain your observations in terms of rotational inertia.

_____

_____

_____

_____

2. Describe in detail the changes in rotational inertia and the rate of rotation of a diver after he leaves the springboard as he performs a one-and-a-half forward-somersault dive. Is the diver's angular momentum conserved throughout the dive?

_____

_____

_____

_____

_____

3. A child stands on a swing and pumps it, setting the swing in motion. Explain why the child and the swing do not represent a closed, isolated system.

_____

_____

_____

10-1

Physics Lab

Pulleys

Objectives

• **Assemble** and **manipulate** a variety of fixed and moveable pulley systems.

• **Calculate** the efficiencies of different pulley systems.

• **Infer** how changing a pulley system affects the ideal mechanical advantage and efficiency of the system.

Pulleys are simple machines that can be used to change the direction of a force, to reduce the force needed to move a load through a distance, or to increase the speed at which the load is moving. Pulleys do not change the amount of work done. However, if the required effort force is reduced, the distance the load moves is decreased in proportion to the distance the force moves. Pulley systems may contain a single pulley or a combination of fixed and movable pulleys.

Procedure

1. Set up the single fixed pulley system, as shown in Figure A1 below.

Figure A

(1)   (2)   (3)   (4)

2. Select a mass that can be measured on your spring scale. Record the value of its mass in Table 1. Determine the weight, in newtons, of the mass to be raised by multiplying its mass in kilograms by the acceleration due to gravity. Recall that $F_g = mg$.

3. Carefully raise the mass by pulling on the spring scale. Measure the height, in meters, that the mass is lifted. Record this value in Table 1. Calculate the work output of the mass by multiplying its weight by the height it was raised. Record this value in Table 2.

4. Using the spring scale, raise the mass to the same height it was raised in step 3. Ask your lab partner to read, directly from the spring scale, the force, in newtons, required to lift the mass. (If your spring scale is calibrated in grams, rather than newtons, calculate the force by multiplying the reading expressed in kilograms by the acceleration due to gravity.) Record this value in Table 1 as the force of

Materials

• 2 single pulleys

• 2 double pulleys

• set of hooked metric masses

• spring scale

• pulley support

• string (2 m)

• meterstick

10-1 Physics Lab

• • • • • • • • • • • • • •

the spring scale. As you are lifting the load with the spring scale, pull upward at a slow, steady rate, using the minimum amount of force necessary to move the load. Any excess force will accelerate the mass and cause an error in your calculations.

5. Measure the distance, in meters, through which the force acted to lift the load to the height it was raised. Record this value in Table 1 as the distance, d, through which the force acts. Determine the work input in raising the mass by multiplying the force reading from the spring scale by the distance through which the force acted. Record the value for the work input in Table 2.

6. Repeat steps 2 through 5 for a different load.

7. Repeat steps 2 through 6 for each of the different pulley arrangements in Figure A. Be sure to include the mass of the lower pulley(s) as part of the mass raised.

8. Count the number of lifting strands of string used to support the weight or load for each arrangement, 1 through 4. Record these values in Table 2.

Data and Observations

Table 1					
Pulley arrange-ment	Mass raised (kg)	Weight (F_g) of mass (N)	Height (h) mass is raised (m)	Force (F) of spring scale (N)	Distance (d) through which force acts (m)
1					
2					
3					
4					

10-1 Physics Lab

Table 2					
Pulley arrange-ment	Work output ($F_g h$) (J)	Work input (Fd) (J)	IMA (d_e/d_r)	Number of lifting strands	Efficiency %
1					
2					
3					
4					

Analysis and Conclusions

1. Find the efficiency of each system. Enter the results in Table 2. What are some possible reasons that the efficiency is never 100%?

   _____

   _____

   _____

2. Calculate the ideal mechanical advantage, IMA, for each arrangement by dividing d_e by d_r. Enter the results in Table 2. What happens to the force, F, as the mechanical advantage gets larger?

   _____

   _____

3. How does increasing the load affect the ideal mechanical advantage and efficiency of a pulley system?

   _____

   _____

4. How does increasing the number of pulleys affect the ideal mechanical advantage and efficiency of a pulley system?

_____

_____

5. The ideal mechanical advantage may also be determined from the number of strands of string supporting the weight or load. How does the calculated *IMA* from question 2 compare with the number of strands of string you counted for each pulley arrangement?

_____

6. Explain why the following statement is false: *A machine reduces the amount of work you have to do.* What does a machine actually do?

_____

_____

_____

_____

Extension and Application

1. In the space provided below, sketch a pulley system that can be used to lift a boat from its trailer to the rafters of a garage, such that the effort force would move a distance of 60 m while the load will move 10 m.

11-1

Physics Lab

Conservation of Energy

In the absence of friction, the work done to pull an object up an inclined plane is equal to the work done to lift it straight up to the same height. The change in gravitational potential energy caused by raising an object is independent of the path through which the object moves.

When an object is lifted straight up, work is done only against gravity. The work equals the increase in the gravitational potential energy of the object, *mgh*. To move the object up a plane, work must be done against gravity and friction. Therefore, to compare the work required to raise the object to the same height but over different paths, you must eliminate the work done against friction from your calculations.

Procedure

1. Weigh the block of wood and enter this value in newtons on the line above Table 1. If your scale measures grams, convert the reading to kilograms and multiply by 9.80 m/s² to find the weight in newtons.

2. Clean and spray with silicone the surfaces of the plane and block to make them as smooth as possible.

Figure A

3. Set the plane at an angle, θ, so that the block will just slide down without being pushed. Measure the incline length, *d*, and the height, *h*, of the high end of the plane, as shown in Figure A. Record these values in Table 1.

Objectives

- **Measure** the forces on an object on an inclined plane.
- **Calculate** the work of lifting an object.
- **Demonstrate** the law of conservation of energy.
- **Compare** theoretical and experimental values for force.

Materials

- inclined plane
- smooth wood block with hook in one end
- spring scale
- meterstick
- string
- silicone spray

11-1 Physics Lab

4. Hook the spring scale to the block and pull the block up the plane at constant velocity. While the block is moving at constant velocity, have your lab partner read the applied force on the scale. Record the value in Table 1 as F_{up}. The force you applied, F_{up}, is the equilibrant of the component of the block's weight parallel to the plane, $F_\parallel$, plus the force of friction, F_f:

$$F_{up} = F_\parallel + F_f.$$

5. Place the block at the top of the incline with the spring scale attached to the hook by a piece of string. Let the block slide down the plane at constant velocity and have your lab partner read and record in Table 1 the force, F_{down}. The frictional force has the same magnitude as before but is exerted in the opposite direction. Thus, the force you exert is the equilibrant of the parallel component of the weight minus the frictional force:

$$F_{down} = F_\parallel + F_f.$$

6. Repeat steps 3–5 twice more, using different angles θ. Keep the height constant in all the trials. Each time you change the angle, measure the length of the plane, d, to the point at which the surface of the plane is at height h from the table, as shown in Figure B. Note that as the angle becomes larger, the length of the plane becomes shorter.

Figure B

Data and Observations

Weight of the block: _____ N

Table 1				
Trial	Length, d (m)	Height, h (m)	F_{up} (N)	F_{down} (N)
1				
2				
3				

11-1 Physics Lab

Table 2				
Trial	$F_{\parallel}$ (N)	Work done without friction, $F_{\parallel}d$ (J)	Potential energy, mgh (J)	Work input, $F_{up}d$ (J)
1				
2				
3				

Analysis and Conclusions

1. Add the two equations for F_{up} and F_{down} from steps 4 and 5 and solve for $F_{\parallel}$.

2. Calculate the value of $F_{\parallel}$ for each trial and enter these values in Table 2.

3. For each trial, calculate and record in Table 2 the work done to pull the block up a frictionless plane. Use the calculated $F_{\parallel}$ and the length of the plane to solve $W = F_{\parallel}d$.

4. Calculate and record the potential energy, *mgh*, of the block at height *h*.

5. Compare the work required to pull the block up the plane set at various angles with the work required to lift the block straight up. Is energy conserved?

_____

_____

_____

Extension and Application

1. Calculate the angle θ for each trial. Using your calculated angle, calculate the theoretical values for $F_\parallel$. Explain any differences between these theoretical values and those in Table 2.

_____

_____

2. A 60.0-kg student skis down an icy, frictionless hill with a vertical drop of 10.0 m. Explain how the law of conservation of energy applies to this situation and calculate how fast she will be going when she reaches the bottom of the hill.

_____

_____

_____

_____

_____

_____

3. Is all of the skier's potential energy in question 2 converted to kinetic energy? Why or why not?

_____

_____

_____

12-1

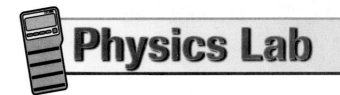

Physics Lab

Specific Heat

Objectives

- **Measure** heat exchange, using a calorimeter.
- **Calculate** the specific heat of metals.
- **Hypothesize** about sources of experimental error.
- **Identify** a material based on its specific heat.

Materials

- CBL unit
- link cable
- temperature probe
- graphing calculator
- glass rod
- string
- safety goggles
- 250-mL beaker
- polystyrene cup
- hot plate
- tap water
- balance
- specific heat set

One property of a substance is the amount of energy that it can absorb per unit mass. This property is called specific heat, C_s. The specific heat is the amount of energy, measured in joules, needed to raise the temperature of 1 kg of a material 1°C (1 K).

A calorimeter is a device that measures the specific heat of a substance. The polystyrene cup, used as a calorimeter, insulates the water-metal system from the environment, while absorbing a negligible amount of heat. Since energy always flows from a hotter object to a cooler one and the total energy of a closed, isolated system always remains constant, the heat energy, Q, lost by one part of the system, is gained by the other:

$$Q_{\text{lost by the metal}} = Q_{\text{gained by the water}}.$$

In this lab, you will determine the specific heat of two different metals. You heat a metal to a known temperature and put it in the calorimeter containing a known mass of water at a measured temperature. You measure the final temperature of the water and material in the calorimeter. Given the specific heat of water (4180 J/kg·K) and the temperature change of the water, you can calculate the heat gained by the water (heat lost by the metal):

$$Q_{\text{gained by the water}} = m_{\text{water}} \Delta T_{\text{water}} (4180 \text{ J/kg·K}).$$

Since the heat lost by the metal is

$$Q_{\text{lost by the metal}} = m_{\text{metal}} \Delta T_{\text{metal}} C_{\text{metal}},$$

the specific heat of the metal can be calculated as follows:

$$C_{\text{metal}} = \frac{Q_{\text{gained by the water}}}{m_{\text{metal}} \Delta T_{\text{metal}}}.$$

Procedure

1. Safety goggles must be worn for this lab. **CAUTION:** *Be careful when handling hot glassware, metals, or hot water.* Fill a 250-mL beaker about half full of water. Place the beaker of water on a hot plate (or a ring stand with a wire screen) and begin heating it.

2. While waiting for the water to boil, measure and record in Table 1 the mass of the metals you are using and the mass of the polystyrene cup.

3. Attach a 30-cm piece of string to each metal sample. Lower one of the metal samples by the string into the boiling water, as shown in Figure A. Leave the metal in the boiling water for at least 5 min.

Hot plate

Figure A

12-1 Physics Lab

4. Connect the temperature probe to the CH1 port of the CBL unit. Attach the graphing calculator to the CBL unit with the unit-to-unit link cable.

5. Start the program PHYSICS on the graphing calculator. From the MAIN MENU, select the option SETUP PROBES. Enter 1 as the number of probes. Then select the TEMPERATURE PROBE from the SELECT PROBE menu. Enter channel number 1 for the port connection.

6. Fill the polystyrene cup half full of room-temperature water. Measure and record in Table 1 the total mass of the water and the cup.

7. Using the temperature probe, reading the temperature on the CBL unit, measure and record in Table 1 the temperature of the boiling water in the beaker and water in the polystyrene cup. Leave the probe in the polystyrene cup water. Do not place the temperature probe onto the hot plate.

8. From the MAIN MENU, select COLLECT DATA. From the DATA COLLECTION menu, select TIME GRAPH. Enter 2.5 for the time between samples. Enter 99 for the number of samples. Press ENTER to continue. If you made an error entering information, use MODIFY SETUP; however, if everything is correct, select USE TIME SETUP.

9. From the TIME GRAPH menu, select LIVE DISPLAY. To set your display range, enter 0 for Ymin, enter 100 for Ymax, and enter 5 for Yscl.

10. Carefully remove the metal from the boiling water and quickly lower it into the water in the polystyrene cup.

11. Press ENTER on the graphing calculator to begin temperature data collection.

12. Gently stir the water in the polystyrene cup for several minutes with the stir rod, keeping the temperature probe from touching the metal.

13. When the CBL unit displays DONE, use the arrow keys to trace the graph. Temperature readings are displayed on the y-axis. Record the highest temperature in your data table. Remove the sample and repeat steps 2, 3, and 6–13.

Data and Observations

Table 1		
	Trial 1	Trial 2
Type of metal		
Mass of calorimeter cup (kg)		
Mass of calorimeter cup and water (kg)		
Mass of metal (kg)		
Initial temperature of room-temperature water (°C)		
Temperature of hot metal (°C)		
Final temperature of metal and water (°C)		

12-1 Physics Lab

• • • • • • • • • • • • • •

Table 2		
	Trial 1	Trial 2
Mass of room-temperature water (kg)		
ΔT metal (°C)		
ΔT room-temperature water (°C)		

Analysis and Conclusions

1. For each trial, calculate the mass of the room-temperature water, the change in temperature of the metal, and the change in temperature of the water in the polystyrene cup. Record these values in Table 2.

2. For each trial, calculate the heat gained by the water (heat lost by the metal).

3. For each trial, calculate the specific heat of the metal. For each metal sample, use the value for heat gained by the water that you calculated in Question 2.

4. For each trial, use the values for specific heat of substances found on page xv to calculate the relative error between your value for specific heat and the accepted value for the metal.

12-1 Physics Lab

5. If you had discrepancies between your values for specific heat of the metal samples and the accepted values, suggest sources of uncertainty in your measurements that may have contributed to the difference.

_____

_____

_____

_____

Extension and Application

1. Obtain a sample of an unknown metal from your teacher. Use the procedure described in this lab to identify it by its specific heat.

2. A 100.0-g sample of a substance is heated to 100.0°C and put in a calorimeter cup (having a negligible amount of heat absorption) containing 150.0 g of water at 25°C. The sample raises the temperature of the water to 32.1°C. Use the values on page xv to identify the substance.

12-2

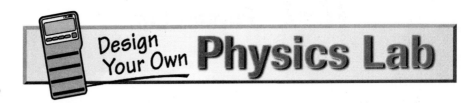

Design Your Own Physics Lab

Graphing the Efficiency of Solar Collectors

When fossil fuels, such as coal and oil, are used up, they cannot be replaced. Harnessing other energy sources becomes increasingly important as nonrenewable resources disappear. Solar power is a promising renewable energy source. Solar collectors gather the sun's energy in a form that people can easily use. In this lab, you will design and build a simple solar collector. You will use a CBL system to measure and compare the heat collected by four cover materials: plastic wrap, wax paper, polyethylene, and glass. Then you can check the cover materials' efficiency.

Objectives

- **Design** a procedure for investigating heat collection.
- **Measure** temperatures and times.
- **Compare** the heat-collecting abilities of four materials.

Possible Materials

- large coffee can with lid
- small coffee can with lid
- sheet of plastic wrap
- sheet of wax paper
- piece of polyethylene
- small pane of glass
- shredded newspapers
- CBL unit
- TI-82 CBL graphics calculator with unit-to-unit link cable
- TI temperature probe
- TI-Graph Link
- colored pencils

Problem

How do four materials compare in their ability to collect and trap heat from solar energy?

Hypothesis

Formulate a hypothesis about the relative efficiencies of plastic wrap, wax paper, polyethylene, and glass in collecting and trapping heat from solar energy.

Plan the Experiment

1. Work in a small group. Decide on a procedure that uses the suggested materials (or others of your choosing) to gather data on how well the four materials (plastic wrap, wax paper, polyethylene, glass) collect heat from sunlight.

2. Decide what kind of data to collect and how to analyze it. You can record your data in the table on the next page. Label the columns appropriately.

3. Write your procedure on another sheet of paper or in your notebook. On the next page, draw the setup you plan to use.

4. **Check the Plan** Have your teacher approve your plan before you proceed with your experiment. Make sure that you understand how to operate the CBL unit, graphics calculator, and temperature probe. Be careful with the electric power source.

12-2 Physics Lab

Setup

Data and Observations

Data Table				

12-2 Physics Lab

Analyze and Conclude

1. **Graphing Data** Use the grid below to graph temperature and time data from your table. How do your plots compare with the STAT PLOT figures on the TI-82?

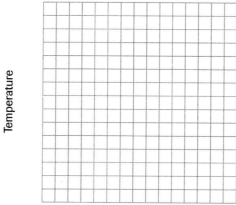

Temperature

Time

_____

_____

_____

_____

_____

_____

_____

_____

_____

2. **Analyzing Data** Which cover material is most efficient at collecting heat? Which is least efficient? Cite your data to support your answers.

_____

_____

_____

_____

_____

_____

3. **Checking Your Hypothesis** Suppose you were stranded on a sunny but cold island in the Arctic Ocean. Luckily you have a supply of glass, polyethylene, wax paper, and plastic wrap and all the tools and structural supports that you need to use them. Which material would you use to build a shelter? Give reasons for your answer.

_____

_____

_____

_____

_____

_____

12-2 Physics Lab

Apply

1. If you wanted to build an inexpensive greenhouse, would you use glass or less-expensive polyethylene in the roof and sides? Why? How could you further improve the efficiency of the structure?

_____

_____

_____

_____

_____

_____

_____

_____

_____

13-1

Physics Lab

Archimedes' Principle

Archimedes' principle states that an object wholly or partially immersed in a fluid is buoyed up by a net force equal to the weight of the fluid that it displaces, $F_{buoyant} = F_{weight\ of\ fluid\ displaced}$. Recall that $F_{weight\ of\ fluid\ displaced} = \rho_{fluid}V_{fluid\ displaced}\ g$, where ρ = density. When an object that is less dense than the fluid is submerged, it sinks only until it displaces a volume of fluid with a weight equal to the weight of the object. At this time, the object is floating partially submerged, equilibrium exists, and $\rho_{fluid}V_{fluid\ displaced} = \rho_{object}V_{object}$.

If an object is more dense than the fluid, an upward buoyant force from the pressure of the fluid will act on the object, but the buoyant force will be too small to balance the downward weight of the denser object. While the object will sink, its apparent weight decreases by an amount equivalent to the buoyant force.

In this experiment, you will investigate the buoyant force of water acting on an object. Recall that 1 mL of water has a mass of 1 g and a weight of 0.01 N. The buoyant force acting on the object is the difference between the weight of the object in air and the weight of the object when it is immersed in water:

$$F_{buoyant} = F_{weight\ of\ object\ in\ air} - F_{weight\ of\ object\ in\ water}.$$

Procedure

1. Pour cool tap water into the 500-mL beaker to the 300-mL mark. Carefully read the volume from the gradations on the beaker and record this value in Table 1.

2. Hang the 500-g mass from the spring scale. Measure the weight of the mass in air and record this value in Table 1.

3. Immerse the 500-g mass, suspended from the spring scale, in the water, as shown in Figure A. Do not let the mass rest on the bottom of the beaker or touch the sides of the beaker. Measure the weight of the immersed mass and record this value in Table 1.

4. Measure the volume of the water with the mass immersed. Record the new volume in Table 1. Remove the 500-g mass from the beaker and set it aside.

Figure A

Objectives

- **Demonstrate** the relationship between the buoyant force and the different weights of an object in air and in water.

- **Calculate** the degree of error between experimentally calculated buoyant force and the weight of the water displaced.

- **Infer** the effect of surface area and density on the amount of water displaced by a known mass.

Materials

- 500-mL beaker
- spring scale
- 500-g hooked mass
- 100-g hooked mass
- polystyrene cup
- paper towel

13-1 Physics Lab

5. Measure and record in Table 2 the volume of water in the beaker. Place the 100-g mass in the beaker of water. Measure and record in Table 2 the volume of the water in the beaker with the mass immersed.

6. The polystyrene cup will serve as a boat. Remove the mass from the water, dry it with a paper towel, and place it in the polystyrene cup. Float the cup in the beaker of water. Measure and record in Table 2 the new volume of water.

Data and Observations

Table 1	
Weight of 500-g mass in air	
Weight of 500-g mass immersed in water	
Volume of water in beaker	
Volume of water in beaker with 500-g mass immersed	

Table 2	
Volume of water in beaker	
Volume of water with 100-g mass immersed	
Volume of water with 100-g mass in polystyrene cup	

13-1 Physics Lab

· · · · · · · · · · · · · ·

Analysis and Conclusions

1. Calculate the buoyant force of water acting on the 500-g mass.

2. Using data from Table 1, calculate the volume of water displaced by the 500-g mass. Calculate the weight of the water displaced. Compare the weight of the volume of water displaced with the buoyant force acting on the immersed object that you calculated in question 1. If the values are different, describe sources of error to account for this difference.

   _____

   _____

3. What happened to the water level in the beaker when the 100-g mass was placed in the poly-styrene cup (boat)? Propose an explanation, which includes density, for any difference in volume you found in steps 5 and 6.

   _____

   _____

   _____

   _____

   _____

   _____

13-1 Physics Lab

Extension and Application

1. Tim and Sally are floating on an inflatable raft in a swimming pool. What happens to the water level in the pool if both fall off the raft and into the water?

_____

_____

_____

_____

_____

2. Icebergs are huge floating masses of ice that have broken from glaciers or polar ice sheets. Only the tip of the iceberg is visible. The density of ice is 0.92 g/cm^3; the density of seawater is 1.03 g/cm^3. What percentage of an iceberg's total volume will be above the surface of the ocean?

14-1

Physics Lab

Ripple Tank Waves

A ripple tank is an ideal medium for observing the behavior of waves. It projects images of waves in the water onto a paper screen below the tank. Light shining through water waves is seen as a pattern of bright lines and dark lines separated by gray areas. The water acts as a series of lenses, bending the light that passes through it. In the ripple tank, the wave crests focus the light from the light source and produce bright lines on the paper. Troughs in the waves cause the light to diverge and produce dark lines. Areas of no disturbance project a uniform gray shade. Dampers placed along the sides of the tank help to prevent reflected waves from interfering with observations.

In this lab, you will investigate wave properties, such as reflection, refraction, diffraction, and interference. The law of reflection states that the angle of incidence is equal to the angle of reflection. Waves undergoing a change in speed and direction as they pass from one medium to another demonstrate refraction. The spreading of waves around the edge of a barrier or a small obstacle exemplifies diffraction. The result of the superposition of two or more waves is interference, which can be either constructive or destructive.

Objectives

- **Observe** reflection, refraction, diffraction, and interference in a ripple tank.
- **Analyze** wave patterns in water.
- **Predict** the behavior of waves in water.

Materials

- ripple tank
- light source
- dowel
- 3 wood blocks
- metal strip
- meterstick
- protractor
- large sheets plain white paper
- eyedropper
- glass plate
- 12 washers
- 2-point-source wave generator
- 0–15 VDC variable power supply

Procedure

Set up the ripple tank as shown in Figure A. Add water to a depth of 5–8 mm. Check the depth at all four corners to be sure the tank is level. Turn on the light source. While creating small disturbances with a pencil, adjust the light source so that clear images of waves appear on the paper screen.

CAUTION: *Do not touch any electric components if your hands are wet.*

Figure A

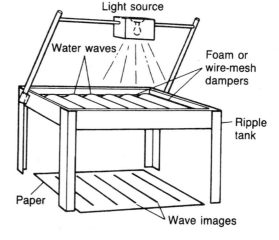

14-1 Physics Lab

· · · · · · · · · · · · · · ·

Reflection

1. Roll the dowel back and forth gently to generate straight pulses at one end of the tank. Place a straight barrier at the opposite end of the tank and send incident pulses toward it so that they strike the barrier head-on (angle of incidence = 0°). Observe and record in item 1 of Table 1 a description of what happens to the pulses as they strike the barrier. Change the angle of the barrier. Generate pulses and record your observations in item 1 of Table 1. Remember that the behavior of pulses is the same as the behavior of waves. For item 2 of Table 1, sketch incoming pulses (waves) striking the barrier at an angle and show the reflected pulses.

2. Use your protractor to measure the angle at which incoming pulses approach the barrier. The angle of incidence, i, is the angle the wave front makes with the normal, N, which is perpendicular to the barrier at the point where the wave meets the barrier. Observe the reflected pulses. The angle between the reflected pulses and the normal is the angle of reflection, r. For item 3 of Table 1, compare the data, the angles of incidence and reflection.

3. Replace the straight barrier with a parabolic strip, arranged so that the open side of the curve is toward the dowel, as shown in Figure B. Send straight pulses toward the curved barrier. Describe in item 4 of Table 1 the shape of the reflected pulses.

Figure B

Parabolic curve

Straight-wave generator

4. Locate the point, or focus, where the reflected pulses come together. Using an eyedropper, let small drops of water fall at that point or, using your finger, gently tap the water at the focus. Note that there are two wave fronts, one that moves outward from the focus and one that moves toward the curved barrier and is reflected. Describe in item 5 of Table 1 the reflection of a circular wave originating at the focus of the concave barrier.

Refraction

1. Place a glass plate in the center of the tank and to one side, as shown in Figure D1 in Table 2. Support the plate with washers, as necessary, to provide an area of shallow water.

2. Send straight pulses toward the glass plate. Carefully observe the pulses at the shallow-water boundary. In Figure D1, sketch the patterns you observe as the waves move through the shallow medium. Rotate the glass plate so that one corner is pointed toward the incident straight waves, as shown in Figure D2. Send a series of straight pulses toward the glass plate. In Figure D2, sketch the patterns you observe as the waves encounter the shallow medium at an angle. Remove the glass plate and washers.

Diffraction

1. Arrange two wood or paraffin blocks, as shown in Figure C. Generate a series of straight pulses and observe the diffraction of the incident waves as they pass through the opening. While generating waves at a constant rate, make the opening progressively smaller. Sketch in item 1 of Table 3 two widths of the incident barrier opening, one large and one small, showing diffraction of the incident waves.

12 cm

Wood or paraffin block

Straight-wave generator

Figure C

2. Increase the rate of wave generation. Recall that the effect of increased frequency is decreased wavelength. Observe what happens when the width of the opening varies. Record your observations in item 2 of Table 3. Remove the blocks from the tank.

14-1 Physics Lab

Name _____

Interference

1. Place the two-point-source generator in the tank near one end. Attach the variable power supply to the generator. Turn the power supply to a low voltage setting so that the point sources produce a continuous series of circular waves of long wavelength. The superposition of waves from the two sources should produce an interference pattern in the tank. Nodal lines, lines of no disturbance where troughs and crests come together simultaneously, should be visible. Antinodal lines, where two troughs or two crests come together simultaneously, should be between the nodal lines. Sketch the wave pattern in item 1 of Table 4. Label the nodal and antinodal lines.

2. Increase the generator frequency and observe how the interference pattern changes. Sketch the new pattern in item 2 of Table 4. Observe what happens to the pattern and number of nodal lines as the generator frequency varies.

Data and Observations

Table 1
1. Observations of straight pulses striking a straight barrier at 0° and at another angle:
2. Sketch of incoming pulses and reflected pulses:
3. Statement of relationship between the angle of incidence and the angle of reflection:
4. Description of the shape of the reflected pulse from a concave surface:
5. Description of the reflection of circular waves originating at the focus:

14-1 Physics Lab

Table 2
Observations of refraction of straight waves:

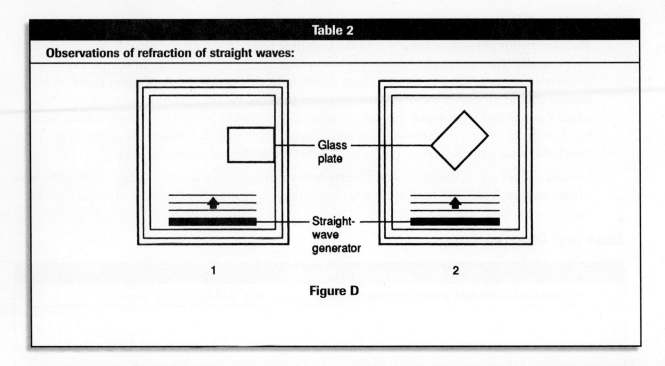

Figure D

Table 3
1. Observations of diffraction:

Large opening	Small opening

2. Observations of diffraction with decreased wavelength:

14-1 Physics Lab

Table 4	
1. Interference pattern:	2. Interference pattern (higher frequency than that used for item 1):

Analysis and Conclusions

1. Using your observations of reflected waves, make a generalization about the angle of incidence and the angle of reflection from a straight barrier.

   _____

   _____

2. What happens to the velocity, wavelength, and frequency of a water wave as it is refracted at the boundary between the deep and shallow water?

   _____

   _____

3. How does wave diffraction change as the width of the opening changes?

   _____

   _____

4. As the wavelength decreases, what happens to wave diffraction?

   _____

   _____

14-1 Physics Lab

5. How does the interference pattern change as the wavelength changes? How does a decrease in wavelength affect the interference pattern?

_____

_____

_____

_____

6. Predict the pattern of waves produced by a double-slit barrier.

_____

_____

7. Put your index finger and thumb together, so that they nearly touch. Hold them in front of a bright, incandescent lightbulb. Look through the tiny opening. Give a reason for the vertical lines you observe.

_____

_____

Extension and Application

1. How can ocean waves help locate underwater reefs or sandbars?

_____

_____

2. How can you apply the law of reflection to sports like platform tennis, racquetball, and pool?

_____

_____

_____

3. Place three wood or paraffin blocks in the tank to create two openings (a double slit), as shown in Figure E. Use the straight-wave generator (dowel) to send waves toward the openings. Sketch your observations. How does this result compare with your prediction in question 6?

Figure E

14-2

Physics Lab

Velocity, Wavelength, and Frequency in Ripple Tanks

Objectives

- **Observe** the behavior of waves in water.
- **Measure** the frequency and length of waves in water.
- **Calculate** wave velocity.
- **Compare** wave velocity under various conditions.

The velocity of all waves of the same kind in any medium is the same. The velocity of a periodic wave can be calculated using

$$v = f\lambda,$$

where v is the velocity of the wave in the medium, f is the frequency of the wave, and λ is the length of the wave. In Lab 14-1, you found that the velocity of a wave changes only when the wave enters a different medium. Since the velocity remains constant, an increase in the frequency results in a decrease in the wavelength. Likewise, a decrease in the frequency results in an increase in the wavelength.

Figure A

Light source

Water waves

Wave generator

Wire mesh dampers

Stroboscope

Ripple tank

To generator power supply

Paper

Materials

- ripple tank
- light source
- point-source wave generator
- 0–15 VDC variable power supply
- stopwatch
- meterstick
- stroboscope
- large sheets plain white paper
- masking tape
- 2 rulers

In this lab, you will measure the frequency and wavelength of water waves and use that information to determine the velocity of the moving waves. Since the ripple-tank images are magnified, measurements of the wavelength in the images must be adjusted for the magnification.

Procedure

1. Set up the ripple tank as shown in Figure A. Add water to a depth of 5–8 mm. Measure the depth at each corner and adjust the tank until it is level. Place the wave generator in the ripple tank so that one point source touches the water. Turn the other point source away from the water. Attach the power supply to the wave generator. Turn

14-2 Physics Lab

on the light source and the power supply for the wave generator. Adjust the power so that the wave generator produces clear images of radiating circles on the paper screen below the ripple tank. Do not change the adjustment on the power supply while doing steps 2–7. Doing so will change your data partway through the activity. **CAUTION:** *Do not operate the generator at a high rate of speed or you may damage it. Do not touch electric components if your hands are wet.*

2. Cover all but every third slit in the stroboscope with masking tape. This should leave four slits open at regular intervals. If not, select a different spacing for your stroboscope so that the intervals between open slits are identical. Hold the stroboscope in one hand with the shaft pointed toward your face. Put a finger of your other hand in the hole of the stroboscope and rotate the strobo-scope disc at a constant rate. Through the stroboscope, observe the wave pattern on the paper screen. Vary the rate of rotation of the stroboscope until the waves appear to stand still.

3. While you are observing the "stopped" waves with the stroboscope, have your partner place the two rulers parallel (along a tangent) to the "stopped" waves and five wavelengths apart on the paper, as shown in Figure B. Since your lab partner sees moving waves, you will have to direct the placement of the rulers. Measure in meters the width of five wavelengths and record this value in Table 1. Calculate the observed wavelength by dividing the distance between the rulers by five. Record in Table 3 the value for one wavelength.

Figure B

4. Measure the length of a ruler, pencil, or block. Place it in the ripple tank and measure the length of the shadow it casts on your paper screen. Record the measurement in Table 2. Find the ratio of the length of the object's shadow to the actual length of the object and record this value in Table 2. This ratio is the magnification factor for the ripple tank. Divide the observed wavelength by the magnification to determine the corrected wavelength and record this value in Table 3.

5. To measure the frequency of the waves, you must determine the stroboscope's rate of rotation. Make a dark mark on one piece of tape on the stroboscope. While you are observing the "stopped" waves with the stroboscope, have your partner watch this mark and measure with the stopwatch the time for ten revolutions. Record the time for ten revolutions in Table 1. Determine the time for one revolution and record this value in Table 1.

6. The frequency of the waves, in hertz, is determined by dividing the number of open slits (which should be four) by the time for one revolution of the stroboscope. Calculate the frequency and record this value in Table 3.

7. Multiply the frequency by the corrected wavelength to determine the velocity. Record this value in Table 3.

8. Repeat steps 2–7 for two trials at different rates of wave generation.

14-2 Physics Lab

Data and Observations

Table 1			
Trial	Width of five wavelengths (m)	Time for ten revolutions (s)	Time for one revolution (s)
1			
2			
3			

Table 2		
Length of object (cm)	Length of shadow (cm)	Magnification ratio (length of shadow/length of object)

Table 3				
Trial	Observed wavelength (m)	Corrected wavelength (m)	Frequency (Hz)	Velocity (m/s)
1				
2				
3				

Physics: Principles and Problems

14-2 Physics Lab

Analysis and Conclusions

1. Which physical quantity did you vary in each trial? Which physical quantity responded to the manipulation?

   _____

   _____

   _____

2. Using your results from Table 3, compare the velocities of the water waves at different frequencies.

   _____

   _____

   _____

3. Does this lab support the rule that the velocity of a wave in a given medium is constant? Give a reason for your answer.

   _____

   _____

   _____

Extension and Application

1. Is the relationship $v = f\lambda$ different from the equation for velocity you used earlier in kinematics, where $v = \Delta d/\Delta t$? Give a reason for your answer.

   _____

   _____

   _____

2. A wave of length 3.5 m has a velocity of 217 m/s. Find the frequency and the period of this wave.

15-1

Physics Lab

The Sound Level of a Portable Radio or Tape Player

Objectives

- **Measure** the sound level of a portable radio or tape player.
- **Determine** the relationship between sound level and volume settings.
- **Recognize** sound levels that may damage hearing.
- **Analyze** sound levels in various situations.

The human ear can sense a wide range of sound intensities. The intensity of a sound at the threshold of pain is 1×10^{12} times greater than that of the faintest detectible sound. However, the ear does not perceive sounds at the threshold of pain as 1×10^{12} times louder than a barely audible tone. Loudness, as measured by the human ear, is not directly proportional to the intensity of a sound wave. Instead, a sound with ten times the intensity of another sound is perceived as twice as loud. A practical unit for measuring the relative intensity of a sound level (β) is the decibel (dB).

$$\beta = (10 \text{ dB}) \log \frac{I}{I_0},$$

where I is the intensity at sound level β and I_0 is a standard reference intensity near the lower limit of human hearing, which corresponds to a sound level of 0 dB. A sound at the threshold of pain corresponds to a sound level of 120 dB.

Table 1 compares intensity ratios and their sound-level equivalents in decibels. Note that when the intensity doubles, the sound level increases by only 3 dB. If the intensity of a sound is multiplied by ten, the sound level increases by 10 dB, but if the intensity is multiplied by 100, the sound level increases by 20 dB.

Table 1	
I/I_0	dB
2	3
3	5
5	7
10	10
20	13
32	15
100	20
1000	30

Table 2	
Sound level (dB)	Exposure per day (h)
90	8
92	6
95	4
97	3
100	2
102	1.5
105	1
110	0.5

15-1

Physics Lab

Prolonged exposure to loud sounds can damage your hearing; the longer the exposure, the greater is the damage. Federal regulations state that workers cannot be exposed to sound levels of more than 90 dB during an 8-h day. Table 2 compares sound levels and the limit of exposure per day to avoid permanent damage to hearing.

In this lab, you will use a sound-level meter, like the one shown in Figure A, to measure the sound level of a portable radio or tape player.

Procedure

1. Check the volume control of your radio or tape player for numerical settings. If it doesn't have them, use white correction fluid to make eight to ten equally spaced marks on the control dial, beginning at the off position.

2. Tune the portable radio to a station that is playing music or insert a cassette tape of your choice into the tape player.

3. Be sure you understand how to use and take readings from the sound-level meter. If you do not, ask your teacher for instructions.

4. Use masking tape to attach the headphones to the sound-level meter, placing the meter microphone against one earpiece.

5. Turn on the radio or tape player and the sound-level meter. Measure the sound level at each increment on the volume control. Record the sound levels in Table 4. Repeat the procedure for other portable radios or players. In Table 4, record these data as additional trials.

6. Turn the volume control to a low setting. Remove the headphones from the sound meter and put them on your head. Tune into some music you enjoy and adjust the volume setting until you have reached the amplitude that you prefer. Record this volume setting in Table 3. Estimate the number of hours, in a given day, that you listen to music at this level. Record this value in Table 3.

Figure A

Materials

- portable radio or tape player, with earphones
- cassette tape of music
- masking tape
- sound-level meter
- white correction fluid
- graph paper

Data and Observations

Table 3	
Volume setting for listening	
Estimated time at preferred volume setting	

15-1 Physics Lab

Table 4							
	Sound level (db)				Sound level (db)		
Setting	Trial 1	Trial 2	Trial 3	Setting	Trial 1	Trial 2	Trial 3
0				6			
1				7			
2				8			
3				9			
4				10			
5							

Analysis and Conclusions

1. Graph sound level (*y*-axis) versus volume setting (*x*-axis). Is there any recognizable relationship between the measured sound level and the corresponding volume setting? Give a reason for your answer.

   _____

   _____

   _____

   _____

2. Use your graph to determine what sound level corresponds to your preferred listening volume setting.

   _____

3. Refer to Table 2 to determine the maximum time you should be listening to music at your preferred volume level. What is this value?

   _____

4. Could your portable radio or tape player be damaging to your hearing? Give a reason for your answer.

   _____

   _____

   _____

   _____

15-1 Physics Lab

Extension and Application

1. Jack's hearing test revealed that he needs a sound level of 20 dB to detect sounds at 2000 Hz. Caroline's hearing test revealed that she needs a sound level of 40 dB to detect sounds at the same frequency. If the normal sound level for testing is 15 dB, how much greater than normal were the sound intensities that Jack and Caroline need to detect sound?

   _____

   _____

   _____

2. The sound level of a rock group is 110 dB, while that of normal conversation at 1 m is 60 dB. Find the ratio of their intensities. Show all your calculations.

3. In open space, the intensity of sound varies inversely with the square of the distance from the source, $I \alpha \frac{1}{r^2}$. Predict what happens to sound levels in an enclosed room. Give a reason for your answer.

   _____

   _____

4. A student is considering a new stereo system for her room and thinks she needs a 150-W amplifier producing 100 W of power. She has asked you if this is a good choice. To help this student, you must find (a) the possible intensity of the sound from this system at a distance $r = 2.5$ m from the source and (b) the corresponding sound level. Use the equation

   $$I = \frac{P}{4\pi r^2},$$

   where P is the power in watts and I is the intensity measured in watts per square meter. $I_0 = 10^{-12}$ W/m^2. Show all your calculations. Is this a wise choice for the student? Give a reason for your answer.

5. If the student in question 4 has one speaker producing 100 dB of sound and she places a second speaker of the same capacity next to it, what is the new sound level at 1 m from the speakers?

   _____

   _____

15-2

Physics Lab

Resonance in an Open Tube

Copyright © by Glencoe/McGraw-Hill

Objectives

- **Measure** the fundamental harmonic and the diameter of an open tube and **relate** it to the sound's wavelength.
- **Calculate** the velocity of sound in air.
- **Form hypotheses** about the difference between an experimental value and a standard.
- **Design** an experiment that establishes resonant lengths in an open tube.

Materials

- adjustable tube, 0.25–1.4 m long
- tuning-fork hammer
- 2 tuning forks, with frequencies between 126 Hz and 512 Hz
- thermometer
- meterstick

Standing longitudinal waves form in air columns in open tubes as well as in closed ones. When a tuning fork vibrates, it disturbs the molecules of air in the tube, so that the air rushes into and out of the tube. A wave reflected at the end of the tube can reinforce or nullify an incident wave. Destructive interference occurs at the nodes. Constructive interference occurs at the antinodes and creates resonance, an increase in the vibrations' amplitude.

At the fundamental harmonic of an open tube, there is a pressure node at each end, with one pressure antinode between. The length, l, of the tube is about one-half of the wavelength, λ, of the sound, or $l = \lambda/2$. The effective length of the air column in a resonating open tube increases by a factor that relates to the diameter, d, of the tube. Thus, the wavelength is

$$\lambda = 2(l + 0.8d).$$

In this experiment, you will apply the principle of resonance to determine the velocity of sound in air. You will use a tuning fork of known frequency, f, and an open tube to find the length at which the air column resonates. From this result, you will use the equation above to find the wavelength of the sound. Then you will use the equation $v = f\lambda$ to find the velocity, v, of sound in air.

To check your results, you will make another calculation. The velocity of sound in air increases as the temperature increases. The velocity of sound in air is 331.5 m/s at 0°C plus 0.59 m/s per degree Celsius above 0°C. You will record the room temperature and determine the accepted value for the velocity of sound in air at that temperature.

Procedure

1. Select an adjustable open tube. Figure A graphs the approximate relationship between the length and the resonant frequency of an open tube of 4.8-cm diameter at room temperature. Using the information in Figure A, select a tuning fork that corresponds to the range of your tube's length.

Open-Tube Length vs Frequency

Figure A

15-2 Physics Lab

2. Record in Table 1 the frequency of your tuning fork.

3. Measure and record in Table 1 the diameter of your adjustable tube.

4. Strike the tuning fork with the tuning-fork hammer. While you hold the tuning fork close to, but not touching, the tube, as shown in Figure B, have your lab partner slowly adjust the length of the tube until you both hear the loudest sound.

Figure B

5. Measure the length of the tube that produces the loudest resonant sound. Record the length in Table 1.

6. The length of the air column must increase by 0.8 times the diameter of the tube to correct for the amount of air vibrating outside the tube. Calculate 0.8d and add this length to the measured length of the tube to determine L, where $L = l + 0.8d$. Record in Table 2 the value of L.

7. Calculate the wavelength and record this value in Table 2.

8. Record in Table 3 the room temperature in °C. Calculate the velocity correction for temperature by multiplying (0.59 m/s/°C)(the room temperature in °C). Enter this correction in Table 3. Calculate the accepted velocity of sound in air by adding 331.5 m/s to the velocity correction. Record the accepted velocity of sound.

9. Repeat steps 4–8, using a tuning fork of different frequency.

Data and Observations

Table 1			
Trial	Frequency (Hz)	Tube diameter (m)	Measured length of tube (m)
1			
2			

15-2 Physics Lab

Table 2		
Trial	Corrected length (*L*) of air column (m)	Wavelength λ = 2*L* (m)
1		
2		

Table 3			
Trial	Air temperature (°C)	Velocity correction for temperature (m/s)	Accepted velocity of sound (m/s)
1			
2			

Analysis and Conclusions

1. Using $v = f\lambda$, determine the velocity of sound in air for each trial.

2. Calculate the relative error between the velocity of sound determined in question 1 and the accepted value from Table 3.

15-2 Physics Lab

3. What effect might steam have on the resonant frequency of the tube?

_____

_____

_____

_____

4. Does an open tube resonate at another length? Test your hypothesis.

_____

_____

_____

Extension and Application

1. An open organ pipe, 3.0 m long and 0.15 m in diameter, resonates when air at 20.0°C is blown against its opening. What is the frequency of the note produced?

2. You can use the difference between the speed of sound and the speed of light to estimate how far away a thunderstorm is. Lightning is visible almost instantaneously. Thunder rumbles travel at about 343 m/s. If you counted 4 s between seeing lightning and hearing thunder, how far away is the storm?

16-1 Physics Lab

Polarized Light

Objectives

- **Observe** light through polarizing filters.
- **Determine** a light's polarity.
- **Describe** applications for polarizing filters.

Materials

- 2 polarizing filters
- Iceland spar calcite crystal
- clear plastic object
- lightbulb with socket and cord
- plane mirror
- safety goggles
- glass plate

While at the beach or swimming pool you probably have noticed the bright reflection of light from the water. If you were wearing polarized sunglasses, they reduced the glare from the reflected sunlight. Light reflected from the water's surface is partially polarized by the reflection effect.

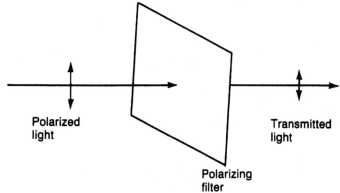

Figure A

Polarized light varies in intensity when it is viewed through a rotating polarizing filter.

Only transverse waves can be polarized, or made to vibrate in one plane. Since light travels in transverse waves, it can be polarized by absorption, reflection, or scattering from shiny objects or materials. Light is also polarized when it passes through certain materials. If you rotate a polarizing filter while observing a light beam, the light intensity may vary from dark to light. Light waves affected in this manner are polarized. If light observed through the rotating filter does not change in intensity, it is unpolarized.

Procedure

1. Hold one polarizing filter between a light source and your eyes. Rotate the polarizing filter through 360°. Observe the intensity of light as you rotate the filter. Record your observations in Table 1.

2. Repeat step 1, adding a second filter. Observe what happens to the intensity of light as you rotate the filter closer to the light source. Record your observations in Table 1.

3. Repeat step 2 and this time rotate the polarizing filter closer to your eye. Record your observations in Table 1.

4. Place an Iceland spar calcite crystal over a written word. Draw what you see in Table 2.

16-1 Physics Lab

5. Hold one polarizing filter over the Iceland spar crystal. Observe the word as in step 4 while you rotate the filter through 360°. Record your observations in Table 2.

6. While rotating one polarizing filter, observe light reflected from a plane mirror. Record your observations of light intensity in Table 3.

7. While rotating one polarizing filter, observe light reflected from a glass plate. Then observe light reflected from a shiny tabletop. Record your observations of light intensity in Table 3.

8. Place a piece of plastic between two polarizing filters. While rotating one of the filters, observe patterns and colors in the plastic. Apply stress to the plastic by pushing on one side. Observe what happens to the patterns and colors as stress is applied and released. Record your observations in Table 4.

Data and Observations

Table 1
Observations of light source with 1 polarizing filter
Observations of light source with 2 polarizing filters, rotating the filter closer to the light
Observations of light source with 2 polarizing filters, rotating the filter closer to your eye

16-1 Physics Lab

Table 2
Observations of a word viewed through Iceland spar
Observations of a word viewed through Iceland spar, with 1 rotating polarizing filter

Table 3
Observations of light reflected from plane mirror
Observations of light reflected from glass plate
Observations of light reflected from shiny tabletop

16-1 Physics Lab

Table 4		
Observations of Plastic	**Unstressed**	**Stressed**

Analysis and Conclusions

1. Is light from a lightbulb polarized? Give a reason for your answer.

_____

_____

_____

_____

2. What happens to the intensity of light viewed through two polarizing filters as one filter rotates? Does it matter which filter rotates? How far must one filter rotate for transmitted light to go from maximum to minimum brightness?

_____

_____

_____

_____

3. Describe the light images viewed through the Iceland spar crystal. What is a possible explanation for this phenomenon?

_____

_____

_____

_____

_____

16-1 Physics Lab

4. Is light reflected from a plane mirror polarized? Give a reason for your answer.

_____

_____

_____

5. Is light reflected from a glass plate or shiny laboratory surface polarized? Give a reason for your answer.

_____

_____

_____

6. Describe the appearance of the unstressed plastic placed between two polarizing filters. What happened as you applied stress to the plastic?

_____

_____

_____

Extension and Applications

1. Describe an orientation of polarizing material on cars that would enable drivers to use their high-beam lights at all times.

_____

_____

_____

_____

2. How might engineers use polarized light and the elastic property of plastic to design better and safer products?

_____

_____

_____

_____

_____

16-1 Physics Lab

3. Use a single polarizing filter to observe light reflected from a glass plate. Examine various angles of reflection until the reflected light is entirely eliminated, as shown in Figure B. With your protractor, measure this angle, θ_p. When unpolarized light strikes a smooth surface and is partially reflected, the light is partially, completely, or not polarized, depending on the angle of incidence. The angle of incidence at which the reflected light is completely polarized is the polarizing angle. The law of reflection states that the angle of incidence and the angle of reflection are equal. An expression relating the polarizing angle, θ_p, to the index of refraction, n, of the reflecting surface is

$$n = \tan \theta_p,$$

which is Brewster's law.

Using the measured angle of reflection, calculate the index of refraction for the glass plate. Substitute a beaker of water and measure the angle of incidence at which reflected light is completely polarized. Calculate the index of refraction of water.

Figure B

16-2

Design Your Own Physics Lab

Intense Light

You probably never give lightbulbs a thought unless one burns out. When you have to replace a burned-out bulb, all you do is read the wattage so you can put in a similar one. But why does the bulb's wattage matter?

Wattage is a measure of electric power, and the brightness, or intensity, of a lightbulb depends on power. In this lab, you will determine the relationship between wattage and intensity. You will measure and compare the intensities of lightbulbs of several different wattages and of different colors.

Problem

How do lightbulbs of different wattage or different color compare in terms of light intensity?

Hypothesis

Formulate a hypothesis about the relationship between intensity and wattage, and about the relative intensity of a colored bulb and a white bulb of the same wattage.

_____

_____

Objectives

- **Measure** the brightness of lightbulbs of different wattages and colors.
- **Interpret** data for intensity and wattage.
- **Determine** the mathematical relationship between intensity and wattage.

Possible Materials

- lamp
- 4 lightbulbs
- colored lightbulb
- thick cotton gloves
- CBL unit with light probe
- TI-82 or TI-83 calculator with unit-to-unit link cable
- L06LIGHT program for calculator
- ring stand and clamp
- books for stacking
- light shield

Plan the Experiment

1. Decide on a procedure that uses the suggested materials (or others of your choosing) to measure the intensity of four white lightbulbs of different wattage and of a colored lightbulb. You should choose a distance of about 0.5 m between the light source and the probe in all cases.

2. Decide what kind of data to collect and how to analyze it. You can record your data in the table on the next page. Label the columns appropriately.

3. Write your procedure on another sheet of paper or in your notebook. On the next page, draw the setup you plan to use.

4. **Check the Plan** Have your teacher approve your plan before you proceed with your experiment. Make sure that you understand how to operate the CBL and light probe. Be careful with the power source. Wear gloves to handle the lightbulbs, which are fragile and can become extremely hot. Let the bulbs cool completely in the lamp before touching them. Wear goggles in case of bulb shattering. Do not look directly at the illuminated lightbulbs.

16-2 Physics Lab

Setup

Data and Observations

Distance between light source and probe: _____ m

Data Table					

16-2 Physics Lab
• • • • • • • • • • • • •

Analyze and Conclude

1. **Graphing Data** Use the grid below to graph average intensity versus average wattage for each bulb tested. Extrapolate (extend) your line graph to the *y*-axis. What is the value of the intensity at the *y*-intercept?

Intensity

Wattage

_____
_____
_____

2. **Analyzing Data** Determine the slope of the graph and write an equation for the line. The equation will relate intensity (*y* in this case) and wattage (*x* in this case). Recall that the equation for a line has the form $y = mx + b$, where *m* is the slope and *b* is the *y*-intercept.

3. **Interpreting Data** Where is the data point for the colored bulb relative to the graphed line? On your graph, measure the vertical distance between the point for the colored bulb and the point on the line that corresponds to the same wattage. What does this distance represent?

_____
_____
_____

4. **Checking Your Hypothesis** Give the coordinates on your graph for the likely data point for a high-intensity, 40-W bulb. Give a reason for your answer.

_____
_____

5. **Checking Your Hypothesis** Describe the graph for nonwhite, same-colored bulbs over the range of wattages used for the white bulbs.

_____
_____
_____

6. **Inferring** What would happen to light intensity if the light source were farther from the light probe? Give a reason for your answer.

_____

_____

_____

_____

_____

7. **Inferring** Why is it unlikely that the *y*-intercept that you extrapolated would be the actual value if a measurement were taken for that point? Would the real graph of intensity versus volume be linear throughout? Give the likely range of wattages for which the equation is useful.

_____

_____

_____

_____

_____

Apply

1. Briefly describe an experiment that would compare the intensities of lightbulbs of different colors but the same wattage. Predict the results of your experiment.

_____

_____

_____

_____

_____

_____

_____

17-1 Physics Lab

Reflection of Light

When a light ray strikes a reflecting surface, the angle of reflection equals the angle of incidence. Both angles are measured from the normal, an imaginary line perpendicular to the surface at the point where the ray is reflected. In this laboratory activity, you will investigate and measure light rays reflected from the smooth, flat surface of a plane mirror in order to determine the apparent location of an image. This type of reflection is called regular reflection. The image of an object viewed in a plane mirror is a virtual image. By tracing the direction of the incident rays of light, you can construct a ray diagram that locates the image formed by a plane mirror.

Objectives

- **Demonstrate** the equality of angle of incidence and angle of reflection.
- **Construct** a ray diagram of light rays reflected in a plane mirror.
- **Calculate** the apparent location and distance of a virtual image from a mirror.

Materials

- thin plane mirror
- small wooden block
- metric ruler
- protractor
- 2 sheets blank paper
- 2 straight pins
- rubber band
- cork board
- thick plane mirror
- 4 thumbtacks
- He-Ne laser

Procedure

A. The Law of Reflection

1. Attach a sheet of paper to the cork board with the tacks or tape. Draw a line, *ML*, across the width of the paper. Attach the small block of wood to the back of the mirror with a rubber band, so that the mirror is perpendicular to the surface of the paper. Center the silvered surface (normally the back side) of the mirror along the line, *ML*, as shown in Figure A.

Figure A

2. About 4 cm in front of the mirror, make a dot on the paper with a pencil and label it *P*. Place a pin, representing the object, upright at point *P*.

3. Place your ruler on the paper about 5 cm to the left of the pin. Sight along the edge of the ruler at the image of the pin in the mirror. When the edge of the ruler is lined up on the image, draw a line along it toward, but not touching, the mirror. Label this line *A*.

4. Move the ruler another 3 or 4 cm to the left and sight along it at the image of the pin in the mirror. Draw a line along the edge of the ruler toward, but not touching, the mirror. Label this line *B*.

17-1 Physics Lab

I (image of pin)

M —————— L

Mirror

r_2 i_2

r_1 i_1

N'

N

P (pin)

Normal

B'

Ruler

Figure B A

5. Remove the pin and the mirror from the paper. Extend lines *A* and *B* to line *ML*. Using dotted lines, extend each of these lines beyond line *ML* until they intersect. Label the point of intersection *I*. This is the position of the image. Measure the object distance from point *P* to line *ML* and the image distance from point *I* to line *ML*. Record these distances in Table 1.

6. Draw a line from point *P* to point *X*, where line *A* meets line *ML*, as shown in Figure B. Using your protractor, construct a normal at this point and measure the angle of incidence, i_1, and the angle of reflection, r_1. Record the values of these angles in Table 1. In a similar way, draw line *PY* from point *P* to point *Y* where line *B* meets line *ML*. Construct the normal at point *Y*. Measure the angle of incidence, i_2, and the angle of reflection, r_2. Record the values of these angles in Table 1. A line such as *PXA* or *PYB* is the path followed by a ray of light from the pin (object) to the mirror, and reflected from the mirror to your eye. Drawing many such rays enables you to locate and recognize images in a mirror.

B. Image Formation

1. Draw a line, *ML*, across the width of another sheet of paper. Draw a triangle in front of the line, as shown in Figure C, and label the vertices *A*, *B*, and *C*. Attach the paper to the cork board and line up the mirror and block on *ML*, as you did in Part A.

2. Place a pin in vertex *A*. Sight twice along the ruler to sketch two lines from the location of the pin at vertex *A* to line *ML*, as you did in Part A. Label these lines A_1 and A_2.

3. Remove the pin from vertex *A* and place it at vertex *B*. Again, sight along the ruler to sketch two lines from the location of the pin at *B* to *ML* and label these lines B_1 and B_2.

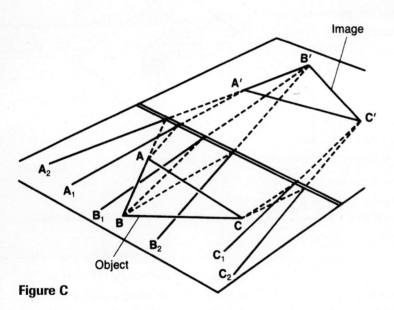

Image

B'

A'

C'

A

A_2

A_1

B_1 B

C

B_2

C_1

Object

C_2

Figure C

17-1 Physics Lab

• • • • • • • • • • • • •

4. Remove the pin from vertex B and place it at vertex C. Repeat the procedure and draw lines C_1 and C_2.

5. Remove the mirror and the pin. Extend the pairs of lines, A_1 and A_2, B_1 and B_2, and C_1 and C_2, beyond the mirror and locate points A', B', and C', as shown in Figure C. Construct the image of the triangle. Measure and record in Table 2 the distances from points A, B, C, A′, B′, and C′ perpendicular to the mirror line ML.

C. Laser Light Reflection

CAUTION: *Avoid staring directly into the laser beam or at bright reflections.*

1. Arrange your thin plane mirror so that the laser beam strikes it at an angle of incidence of about 60° from the normal. Reflect the laser beam onto the ceiling. In Table 3, draw your observations of the reflected image. Repeat the process, using a clean, thick plane mirror (rear surfaced). In Table 3, record your observations of the reflected image.

2. With the room lights off, spray a fine mist of water in the air (or clap two chalk-filled erasers) over the reflected beam. In Table 3, record your observations.

Data and Observations

Table 1	
Object distance	
Image distance	
Angle of incidence, i_1, PXN	
Angle of reflection, r_1, AXN	
Angle of incidence, i_2, PYN′	
Angle of reflection, r_2, BYN′	

Table 2	
Point	Distance to Mirror Line (cm)
A	
B	
C	
A′	
B′	
C′	

17-1 Physics Lab

Table 3	
Drawing of reflected laser light from thin plane mirror	
Drawing of reflected laser light from thick plane mirror	
Observation of reflected laser beams	

Analysis and Conclusions

1. Using your observations from Table 1, what can you conclude about the angle of incidence and the angle of reflection?

   _____

2. How far behind a plane mirror is the image of an object in front of the mirror?

   _____

   _____

3. Using your observations from Table 2, compare the size and orientation of your constructed image with those of the object triangle.

   _____

   _____

   _____

17-1 Physics Lab

· · · · · · · · · · · · · ·

4. Using your observations in this experiment, summarize the general characteristics of images formed by plane mirrors.

_____

_____

_____

5. Why do you think the image formed by a plane mirror is called virtual rather than real?

_____

_____

_____

6. When you used the thick plane mirror, which reflected spot of the laser beam was brightest? Give an explanation for the bright beam.

_____

_____

_____

7. Explain the difference in patterns of the reflected laser beam from the thin and thick plane mirrors. Draw a diagram that demonstrates the pattern of laser light reflected from a thick plane mirror and your explanation for this result.

_____

_____

_____

_____

Extension and Application

1. Use small blocks of wood or a stand to set a piece of plate glass vertically on a line, *ML*, drawn on a sheet of paper that is attached to a cork board. The glass must be perpendicular to the paper, and you must be able to sight through the middle portion of the glass. Place one pin upright in front of the plate glass. Place a second pin upright in back of the plate glass at a point where the reflected image from the plate glass seems to be, as you sight through the glass. View the image at different angles and adjust the second pin slightly until this pin behind the plate glass is always at the position of the image, regardless of the viewing angle. Measure the distance from the object pin to line *ML* and compare it to the distance from the second pin to *ML*.

_____

_____

2. Place a mirror vertically on a flat surface as you did in Problem 1. Place a small card with some writing on it about 5 cm in front of the edge of the mirror that is closest to you. Orient the card so that the writing is upside down when you try to read it. Look into the mirror at an angle of 45°. Report what you see. Place a second mirror next to the first one with their edges touching at an angle of 90°. Look into both mirrors at the junction, so that you see each mirror at an angle of 45°. Examine the image carefully. Report what you observe. Explain your observations.

    _____

    _____

    _____

    _____

    _____

3. Many department stores have large plane mirrors high on walls or on projections from ceilings. These may be one-way mirrors that allow surveillance of the store. From one side, this surface looks like a mirror, but from the other side, the activities of the shoppers can be observed. How does this type of mirror work?

    _____

    _____

    _____

    _____

    _____

    _____

17-2

Physics Lab

Snell's Law

Objectives

- **Measure** the angles of incidence and refraction for light passing through glass.
- **Calculate** the index of refraction, using Snell's law.
- **Compare** the entrance and exit angles of refraction for glass.

Materials

- glass plate
- metric ruler
- protractor
- 1 sheet of paper
- assortment of unknown materials

Light travels at different speeds in different media. As light rays pass at an angle from one medium to another, they refract or bend at the boundary between the two media. If a light ray enters an optically more dense medium at an angle, it bends toward the normal. If a light ray enters an optically less dense medium, it bends away from the normal. This change in direction of light at the boundary of two media is called refraction.

The index of refraction of a substance, n_s, is the ratio of the speed of light in a vacuum, c, to its speed in the substance, v_s:

$$n_s = \frac{c}{v_s}.$$

All indices of refraction are greater than one, because light always travels slower in media than in a vacuum.

The index of refraction is also obtained from Snell's law, which states that a ray of light bends in such a way that the ratio of the sine of the angle of incidence to the sine of the angle of refraction is a constant. Snell's law can be written

$$n = \frac{\sin \theta_i}{\sin \theta_r}.$$

For any light ray traveling between media, Snell's law, in a more general form, can be written

$$n_i \sin \theta_i = n_r \sin \theta_{r'}$$

where n_i is the index of refraction of the incident medium and n_r is the index of refraction of the second medium. The angle of incidence is θ_i, and the angle of refraction is θ_r.

Procedure

1. Place the glass plate in the center of a sheet of paper. Use a pencil to trace an outline of the plate.

2. Remove the glass plate and construct a normal N_1B at the top left of the outline, as shown in Figure A.

Figure A

Normals N_1 and N_2 to the surface of the glass at C and D, respectively

17-2 Physics Lab

3. Use your ruler and protractor to draw a heavy line AB at an angle of 30° with the normal. Angle ABN_1 is the angle of incidence, θ_i.

4. Replace the glass plate in the outline on the paper. With your eyes on a level with the glass plate, sight along the edge of the glass plate opposite the line AB until you locate the heavy line through the glass, as shown in Figure B. Sight your ruler at the line until its edge appears to be a continuation of the line. Draw the line CD as shown in Figure A.

Figure B

5. Remove the glass plate and draw another line CB, connecting lines CD and AB. Extend the normal N_1B through the rectangle, forming a new line $N_1BN_{1'}$.

6. Use a protractor to measure angle $CBN_{1'}$. This is the angle of refraction, θ_r. Record the value of this angle in Table 1. Consult the table of trigonometric functions in Appendix A to find the sines corresponding to the measured angles of incidence and refraction. Record these values in Table 1. Determine the ratio of $\sin \theta_i$ to $\sin \theta_r$ and record this value in the table as the index of refraction, n.

7. Construct a normal N_2 at point C. Measure angle DCN_2, which will be called $\theta_{r'}$, and record this value in the table.

8. Turn the paper over and repeat steps 1 through 7, using an angle of incidence of 45°. Record all the data in Table 1. Again, determine the index of refraction from your data.

Data and Observations

Table 1					
θ_i	θ_r	$\sin \theta_i$	$\sin \theta_r$	$\theta_{r'}$	Index of refraction, n
30°					
45°					

17-2 Physics Lab

Analysis and Conclusions

1. Is there good agreement between the two values for the index of refraction of plate glass?

   _____

   _____

   _____

   _____

2. According to your diagrams, are light rays refracted away from or toward the normal as they pass at an angle from an optically less dense medium into an optically more dense medium?

   _____

   _____

   _____

3. According to your diagrams, are light rays refracted away from or toward the normal as they pass from an optically more dense medium into an optically less dense medium?

   _____

   _____

   _____

4. Compare θ_i and $\theta_{r'}$. Is the measure of $\theta_{r'}$ what you should expect? Give a reason for your answer.

   _____

   _____

   _____

   _____

5. The speed of light in air is so close to the speed of light in a vacuum that you can consider them to be equal when answering this question. Use your results to determine the approximate speed of light as it travels through glass. By what percentage is the speed of light traveling in a vacuum faster than the speed of light traveling in glass?

   _____

   _____

   _____

   _____

   _____

17-2 Physics Lab

Extension and Application

1. Will light be refracted more while passing from air into water or while passing from water into glass? Give a reason for your answer.

_____

_____

_____

_____

_____

2. Gemologists can identify an unknown gem by measuring its index of refraction, specific to each type of gemstone. The gemologist places the unknown gem in a series of liquids with known indices of refraction. When the gem becomes nearly invisible in a liquid, the gemologist concludes that the gem has an index of refraction corresponding to that of the liquid in which it is immersed. With this information, the gemologist consults an index of refraction of gemstones to identify the unknown gem. Table 2 lists the indices of refraction of some liquids and materials.

Table 2			
Liquid	**n**	**Material**	**n**
Water	1.34	Glass	1.48–1.7
Olive oil	1.47	Quartz	1.54
Mineral oil	1.48	Beryl	1.58
Oil of wintergreen	1.48	Topaz	1.62
Clove oil	1.54	Tourmaline	1.63
Cinnamon oil	1.60	Garnet	1.75

Use these liquids to identify the unknown materials your teacher gives you. Use tweezers to carefully place each sample into each bottle of liquid. Observe the material and the behavior of light as it passes through the liquid and the material. Before immersing the sample in the next bottle of liquid, rinse it off in water and dry it gently to prevent contamination of the oils. When the material seems to disappear, record the index of refraction for the liquid and use this value to identify the material.

_____

_____

_____

_____

_____

18-1

Physics Lab

Concave and Convex Mirrors

Objectives

- **Demonstrate** image formation with concave and convex spherical mirrors.

- **Observe** the properties of images formed by spherical mirrors.

- **Measure** the focal length of spherical mirrors.

- **Calculate** the focal length of a spherical mirror, using the mirror equation.

Spherical mirrors are portions of spheres and have one silvered side that is a reflecting surface. If the inner side is the reflecting surface, the mirror is concave. If the outer side is the reflecting surface, the mirror is convex.

The center of the sphere of which the mirror is a portion is the center of curvature (C) of the mirror. An imaginary line that is perpendicular to the center of the mirror (A) and that passes through C is the principal axis of the mirror. The point halfway between C and A is the focal point (F) of the mirror. The distance from F to A is the focal length (f) of the mirror. The distance of the object from the mirror, d_o, and the distance of the image from the mirror, d_i, are related to the focal length by the lens/mirror equation

$$\frac{1}{f} = \frac{1}{d_i} + \frac{1}{d_o}.$$

Figure A illustrates these relationships.

Concave mirrors produce real images, virtual images, or no image, depending upon how far the object is from the mirror. Real images can be captured on a screen, and light rays actually pass through them. Convex mirrors produce only virtual images. Virtual images cannot be captured on a screen and appear to originate behind the mirror.

Light rays that approach a concave mirror and are parallel to the principal axis reflect and converge at the focal point. If all the light rays approaching a concave mirror are parallel to the principal axis, they will meet at or near the focal point and form an image of the object.

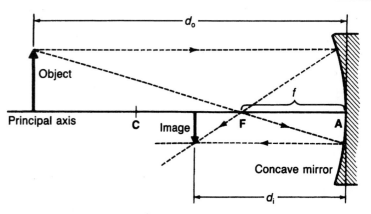

Figure A

The center of the curvature C is located at a distance of twice the focal length f from the center of the mirror, along the principal axis.

18-1 Physics Lab

Materials

concave mirror

convex mirror

2 metersticks

cardboard screen

holders for mirror, screen, and light source

masking tape

metric ruler

light source

Procedure

A. Focal Length of a Concave Mirror

Arrange your mirror, metersticks and screen as shown in Figure B. If the sun is visible, the focal length can be determined by projecting a focused image of the sun onto a screen and measuring the distance from the mirror's center to the screen. **CAUTION:** *Do not look directly at the sun or you may severely damage your eyes.* Another method is to point the mirror at a distant object (more than 10 m away) and move the screen along the meterstick until you obtain a sharp image of the distant object on the screen. The distance between the mirror and the screen is the approximate focal length of the mirror. Record in Table 1 the focal length of your mirror.

Figure B

Finding the focal length of a mirror

Figure C

B. Concave Mirror

1. The center of curvature of the mirror, C, is twice the focal length. Record this value in Table 1. Arrange the two metersticks, mirror, light source, and screen as shown in Figure C. Use masking tape to hold the metersticks in place.

2. Place the light source at a distance greater than C from the mirror. Measure the height of the light source and record this value in Table 1. Move the screen back and forth along the meterstick until you obtain a sharp image of the light source. Determine the distance of the image from the mirror, d_i, by measuring from the mirror's center to the screen. Record in Table 2 your measurements of d_i, d_o, and h_i and your observations of the image.

•••••••••••••••

3. Move the light source to *C*. Move the screen back and forth until you obtain a sharp image. Record in Table 2 your measurements of d_i, d_o, and h_i and your observations of the image.

4. Move the light source to a position between *F* and *C*. Move the screen back and forth until you obtain a sharp image. Record in Table 2 your measurement of d_i, d_o, and h_i and your observations of the image.

5. Move the light source to a distance *F* from the mirror. Try to locate an image on the screen. Observe the light source in the mirror. Record your observations in Table 2.

6. Move the light source to a position between *F* and *A*. Try to locate an image on the screen. Observe the image in the mirror. Record your observations in Table 2.

C. Convex Mirrors

Place the convex mirror in the holder. Place the light source anywhere along the meterstick. Try to obtain an image on the screen. Observe the image in the mirror. Move the light source to two other positions along the meterstick and try each time to produce an image on the screen. Observe the image in the mirror. Record your observations in Table 3.

Data and Observations

Table 1	
Focal length of mirror, *f*	
Center of curvature of mirror, *C*	
Height of light source, h_o	

18-1 Physics Lab

Table 2					
Position of object	Beyond C	At C	Between C and F	At F	Between F and A
d_o					
d_i					
h_i					
Type of image: real, none, or virtual					
Direction of image: inverted or erect					

Table 3					
Trial	Position of object	Position of image	Type of image: real or virtual	Image size compared to object size	Direction of image: inverted or erect
1					
2					
3					

18-1 Physics Lab

Analysis and Conclusions

1. Use your observations from Table 2 to summarize the characteristics of images formed by concave mirrors in each situation.

   a. The object is beyond the center of curvature.

   _____

   _____

   b. The object is at the center of curvature.

   _____

   _____

   c. The object is between the center of curvature and the focal point.

   _____

   _____

   d. The object is at the focal point.

   _____

   _____

   e. The object is between the focal point and the mirror.

   _____

   _____

2. Use your observations from Table 3 to summarize the characteristics of images formed by convex mirrors.

   _____

   _____

3. For each of the real images you observed, use the lens/mirror equation to calculate f. Do your calculated values agree with each other?

● ● ● ● ● ● ● ● ● ● ● ●

4. Average the values of *f* that you calculated for question 3 and compute the relative error between the average and the measured value for *f* recorded in Table 1.

socket

**Illuminated
lightbulb**

Figure D

Illusion apparatus

Extension and Application

1. The image of an illuminated lightbulb in the apparatus in Figure D can be projected onto the empty socket on top of the box. Describe how you would orient a large, spherical mirror to reflect the illuminated lightbulb in the box onto the socket on top.

_____

_____

_____

_____

_____

18-2

Physics Lab

Convex and Concave Lenses

Objectives

- **Demonstrate** the formation of images from concave and convex lenses.

- **Define** the characteristics of the images formed by concave and convex lenses.

- **Analyze** image size versus distance for a concave lens.

- **Interpret** a graph to find the focal length of a concave lens.

A convex, or converging, lens is thicker in the middle than at the edges. A concave, or diverging, lens is thinner in the middle than at the edges. The principal axis of the lens is an imaginary line perpendicular to the plane that passes through the lens's midpoint. It extends from both sides of the lens. At some distance from the lens along the principal axis is the focal point (F) of the lens. Light rays that strike a convex lens parallel to the principal axis come together or converge at this point. The focal length of the lens depends on both the shape and the index of refraction of the lens material. As with mirrors, an important point is located at a distance twice the focal length. If the lens is symmetrical, the points F and 2F are the same distance on both sides of the lens, as shown in Figure A.

Figure A

Focal length, object distance, and image distance are measured from the lens along the principal axis.

A concave lens causes all incident parallel light rays to diverge. Rays approaching a concave lens parallel to the principal axis appear to intersect on the near side of the lens. Thus, the focal length of a concave lens is negative. Figure B shows the relationship between the incoming and refracted rays passing through a concave lens.

The distance from the center of the lens to the object is d_o, and the distance from the center of the lens to the image is d_i. The lens/mirror equation is

$$\frac{1}{f} = \frac{1}{d_i} + \frac{1}{d_o}.$$

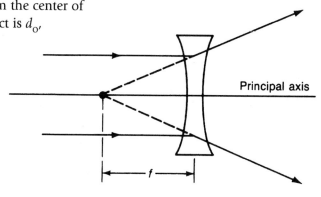

Figure B

The focal length of a concave lens is negative. All light rays that pass through a concave lens diverge.

18-2 Physics Lab

In this activity, you will measure the focal length, f, of a convex lens and place an object at various distances from the lens to observe the location, size, and orientation of the images. You will find the focal length of a concave lens by tracing diverging rays backward to a point of intersection. Recall that real images can be projected onto a screen; virtual images cannot.

Materials

double convex lens

concave lens

meterstick

meterstick support

small cardboard screen

light source

holders for screen, light
    source, and lens

metric ruler

sunlight

Procedure

A. Focal Length of a Convex Lens

To find the focal length of the convex lens, arrange your lens, meterstick, and screen as shown in Figure C. Point the lens at a distant object and move the screen back and forth until you obtain a clear, sharp image of the object on the screen. A darkened room makes the image easier to see. Record in Table 1 your measurement of the focal length. Compute the distance 2F and record this value in Table 1.

Figure C

B. Convex Lens

1. Arrange the apparatus as shown in Figure D. Place the light source somewhere beyond $2F$ on one side of the lens and place the screen on the opposite side of the lens. Move the screen back and forth until a clear, sharp image forms on the screen. Record in Table 1 the height of the light source (object), h_o. Record in Table 2 your measurements of d_o, d_i, and h_i and your observations of the image.

2. Move the light source to $2F$. Move the screen back and forth until a clear, sharp image forms on the screen. Record in Table 2 your measurements of d_o, d_i, and h_i and your observations of the image.

Light source

Convex lens
in lens holder

Screen

Figure D

Copyright © by Glencoe/McGraw-Hill

18-2 Physics Lab

• • • • • • • • • • • • • •

3. Move the light source to a location between F and $2F$. Move the screen back and forth until a clear, sharp image forms on the screen. Record in Table 2 your measurements of d_o, d_i, and h_i and your observations of the image.

4. Move the light source to a distance F from the lens. Try to locate an image on the screen. Record your observations in Table 2.

5. Move the light source to a position between F and the lens. Try to locate an image on the screen. Look through the lens at the light source and observe the image. Record your observations in Table 2.

C. Concave Lens

Place a concave lens in a lens holder and set it on the meterstick. Place a screen on one side of the lens. Use one of the following procedures to determine the focal length.

Sunlight: Allow the parallel rays of the sun to strike the lens along the principal axis so that an image forms on the screen. **CAUTION:** *Do not look directly at the sun or you may severely damage your eyes.* The image should be a dark circle inside a larger, brighter circle. Place the screen close to the lens. Quickly measure the distance from the lens to the screen and the diameter of the bright circle. Record these data in Table 3. Move the screen to five other positions and repeat the measurements.

He-Ne laser: Shine a laser beam through the concave lens so that an image forms on the screen. **CAUTION:** *Do not look directly at the laser source or you may severely damage your eyes.* Measure the distance from the screen to the lens and the diameter of the circle of light projected onto the screen. Record these data in Table 3. Move the screen to five other positions and repeat the measurements.

Data and Observations

Table 1	
Focal length, f	
2F	
Height of light source, h_o	

18-2 Physics Lab

Name _____

Table 2					
Position of object	Beyond 2F (cm)	At 2F (cm)	Between 2F and F (cm)	At F (cm)	Between F and lens (cm)
d_o					
d_i					
h_i					
Type of image: real, none, or virtual					
Direction of image: inverted or erect					

Table 3	
Distance from lens (m)	Diameter of screen image (cm)

18-2 Physics Lab

Analysis and Conclusions

1. Use the data from Table 2 to summarize the characteristics of images formed by convex lenses in each situation.

   a. The object is beyond 2F.

   _____

   _____

   b. The object is at 2F.

   _____

   _____

   c. The object is between 2F and F.

   _____

   _____

   d. The object is at F.

   _____

   _____

   e. The object is between F and the lens.

   _____

   _____

2. For each of the real images you observed, calculate the focal length of the lens, using the lens/mirror equation. Do your values agree with each other?

3. Average the values for f found in question 2 and calculate the relative error between this average and the value for f from Table 1.

Name _____

4. Use your data in Table 3 to plot a graph on a sheet of graph paper of the image diameter versus the distance from the lens. Place the image diameter on the vertical axis and the distance from the lens on the horizontal axis. Allow room along the horizontal axis for negative distances. The rays expanding from the lens appear to originate from the focal point. Draw a smooth line that best connects the data points and extend the line until it intersects the horizontal axis. The negative distance along the horizontal axis at the intersection represents the value of the focal length. What is the focal length you derived from your graph? If your lens package includes an accepted focal length, calculate the relative error for the focal length you determined. Use your graph paper for calculations.

Extension and Application

1. Use two identical, clear watch glasses and a small fish aquarium to investigate an air lens. Carefully glue the edges of the watch glasses together with epoxy cement or a silicone sealant so that the unit is watertight. Attach the air lens to the bottom of an empty fish aquarium with a lump of clay and put an object nearby, as shown in Figure E. Observe the object through the lens and record your observations. Predict what will happen when the aquarium is filled with water and light passes from a more dense medium to a less dense one as it passes through the air lens. Fill the aquarium with water and repeat your observations. Compare the two sets of observations. Explain your results. What would you expect if you used a concave air lens? Design and construct a concave air lens to test your hypothesis.

_____

_____

_____

_____

_____

_____

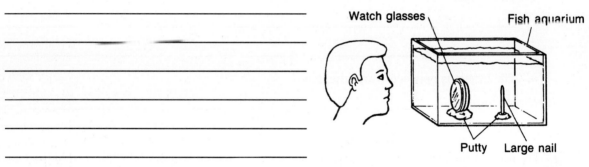

Figure E

2. The movement of reflected laser light relative to an observer's moving head can be used to determine nearsightedness or farsightedness. Observers should remove their glasses or contact lenses. In a darkened room, use a concave lens to expand the beam diameter of a laser to project a large spot on a screen. Observers should move their heads from side to side while looking at the spot. Each student should record the direction in which the reflected speckles of laser light appear to move and the direction in which his or her head is moving. Then observers should replace their glasses or contact lenses and repeat their observations.

_____

_____

_____

_____

_____

_____

19-1

Physics Lab

Double-Slit Interference

Objectives

- **Measure** distances between interference lines as accurately as possible.
- **Calculate** the wavelength of colored light.
- **Predict** the locations of first-order lines for other colors of light.

As Thomas Young demonstrated in 1801, light falling on two closely spaced slits passes through the slits and is diffracted. The spreading light from the two slits overlaps and produces an interference pattern, an alternating sequence of dark and light lines, on a screen. A bright band appears at the center of the screen. On either side of the central band, alternating bright and dark lines appear on the screen. Bright lines appear where constructive interference occurs, and dark lines appear where destructive interference occurs. The first bright line on either side of the central band is the first-order line. This line is bright because, at this point on the screen, the path lengths of light waves arriving from each slit differ by one wavelength. The next line is the second-order line. Several lines are visible on either side of the central bright band. The equation that relates the wavelength of light, the separation of the slits, the distance from the central band to the first-order line, and the distance to the screen is

$$\lambda = xd/L.$$

Figure A shows the variables in this relationship.

Figure A

Materials

- double slit
- clear filament light with socket
- red, yellow, and violet transparent filter
- rubber band
- meterstick
- metric ruler
- magnifier with scale
- masking tape
- index cards, 12 × 18 cm

Procedure

CAUTION: *Avoid staring directly into the laser beam or at bright reflections.*

1. Measure the slit separation of your double slit, using the magnifier and a graduated millimeter scale. If it is known, use the actual slit separation. Record in millimeters the separation, *d*, in Table 2. This value is constant for all trials.

2. Place a red filter over the lightbulb and secure it with a rubber band. Turn on the lightbulb. Turn off the classroom lights.

3. Stand 1 m from the light and observe the lightbulb through the double slit. Record in Table 1 your observations of the light bands. Move several meters away from the light and repeat your observations, recording them in Table 1.

19-1 Physics Lab

4. Move approximately 2–4 m from the light and mark the location with a small piece of tape. Measure the distance from the lightbulb filament to the small piece of tape. Record the distance, L, in Table 2.

5. Make a dark line on an index card. With your lab partner viewing the light through the double slit, align the card, as directed by your partner, so that the dark line is located over the central bright band, or lightbulb filament.

6. While your lab partner observes the lightbulb through the double slit, slowly move your pencil back and forth until your partner tells you that it is over one of the first-order lines. Make a mark on the index card where the first-order line occurs. Measure the distance, x, from the central bright line to the first-order line. Record this value in Table 2.

7. Remove the red filter from the light. Repeat steps 4–6 with the other colored filters. Record your data in Table 2.

Data and Observations

Table 1	
	Observations of Light
Close	
Far away	

Table 2			
Color	**d (mm)**	**x (m)**	**L (m)**
Red			
Yellow			
Violet			

19-1 Physics Lab

Analysis and Conclusions

1. Describe how the double-slit pattern changes as you move farther from the light source.

   _____

   _____

2. Calculate the wavelength, in nm, for each of the colors. Show your work.

3. Which color of light will be diffracted most as it passes through the double slit? Give a reason for your answer.

   _____

   _____

   _____

4. Predict the location of the first-order line for infrared light of wavelength 1.000×10^3 nm. How does this location compare with the other first-order lines observed in this experiment?

   _____

   _____

   _____

5. Predict the location of the first-order line for ultraviolet light of wavelength 375 nm. How does this location compare with the other first-order lines observed in this experiment?

   _____

   _____

   _____

19-1 Physics Lab

Extension and Application

1. Place a double slit over the front of a He-Ne laser or a laser pointer. Project the interference pattern onto a screen. **CAUTION:** *Avoid staring directly into the laser beam or at bright reflections.* Measure L and x for the first-order line. Calculate the separation of the double slit, using the value 632.8 nm for the wavelength of the He-Ne laser light or 670 nm for laser-pointer light.

2. A bright light, visible from 50.0 m away, produces an angle of 0.8° to the first-order line when viewed through two slits separated by 0.05 mm. What color is the light?

3. Blue-green light of wavelength 5.00×10^2 nm falls on two slits that are 0.0150 mm apart. A first-order line appears 14.7 mm from the central bright line. What is the distance between the slits and the screen? Show your calculations.

4. The slit separation and the distance between the slits and the screen are the same as those in Question 3, but sodium light of wavelength 589 nm is used. What is the distance from the central line to the first-order line with this light? Show your calculations.

19-2

Physics Lab

White-Light Holograms

A unique application of lasers is the production of holograms. Holograms are three dimensional images or photographs produced by interference of coherent light. Two primary types of holograms are reflection and transmission. Most of you have seen a white-light, reflection hologram, such as those on cereal boxes, on credit cards, and in magazines. Reflection holograms are created when laser light goes through the film and reflects off the object back onto the film. The incident light and the reflected light create an interference pattern on the film.

Holograms with greater depth and clarity are transmission holograms. They are produced when a laser beam is split into two parts. One part of the beam reflects off the object being photographed and strikes the holographic film, while the other falls directly on the film. The two superimposed beams form an interference pattern, which is recorded on the film. The film picks up the intensity of the incident light and its phase relationship. When the developed film is viewed, it is illuminated by a coherent light beam of the same wavelength used to expose the film, and the viewer looks back at the transmission hologram along the direction of the beam's origination.

Procedure

1. Follow your teacher's safety instructions for using photographic chemicals to develop your holographic image. **CAUTION:** *The compounds used to develop the film are poisonous and caustic. If you use any of these compounds, protect your eyes, skin, and clothing.*

2. To construct a vibration-isolation system, partially inflate the inner tube and place it on a stable lab table. Set the steel plate (or another massive piece of equipment) on the inner tube. This setup will minimize vibrations that can affect the quality of the hologram while the film is exposed.

3. Place the small concave mirror in the helping hand. Place the helping hand on the steel plate, as shown in Figure A.

4. Place the laser on an adjacent table and turn off the room lights. Direct the beam onto the mirror. **CAUTION:** *Avoid staring directly into the laser beam or at bright reflections.* The reflected, expanded beam from the laser should be directed toward the steel plate. Place the small green light nearby so that it is just possible to see the steel plate in the darkened room. Turn the room lights back on.

5. Trim a small index card to the size of your holographic film plate (a 6.5-cm square). Arrange the small disk magnets on the steel plate so that four columns support each corner of the trimmed index card

Objectives

- **Demonstrate** the creation of a hologram from the interference of light waves.
- **Compare** a holographic image to an ordinary photograph.
- **Observe** the generation of colors from the interference pattern of white light.

Materials

- He-Ne Laser
- small concave mirror
- holographic film plates
- automobile inner tube
- bicycle air pump
- 2 index cards, 12 × 18 cm
- holographic developer and bleach solution
- plastic tongs
- rubber gloves
- green safelight
- 3 plastic trays
- quarter
- steel plate
- helping-hand project holder
- 12 disk magnets
- stopwatch
- black paper

19-2 Physics Lab

Mirror

Laser beam

Helping hand

Expanding beam

Disk magnets

Small pile of coins or shiny objects

Steel plate

Figure A

(temporarily taking the place of the glass plate). Adjust the height of the magnet supports so that the index card is just above the objects being photographed. Adjust the helping hand so that the reflected beam illuminates the index card. Remove the index card.

6. Select one or more metallic, shiny objects. Arrange them within the illuminated area. Block the laser beam with a 12 × 18 cm index card. The index card will serve as a camera shutter.

7. Turn off the room lights. Carefully open the film container and remove one holographic film plate or one piece of film. Hold the glass plate by its edges or hold the film carefully by a corner. Do not touch the center of the plate or film. Close and reseal the remaining plates or film in the light-tight container. Place the glass plate on the magnet supports, or put the film between two glass plates and place this assembly on the magnet supports. The emulsion side of the plate or film feels slightly sticky and has a dull appearance. Make sure that the emulsion side of the plate or film is facing the objects.

8. During exposure of the hologram, all students in the room should minimize movement to help reduce vibrations. While one lab partner removes the index-card shutter, the other partner should begin timing the exposure. Try an exposure time of 20 seconds or follow your teacher's instructions for timing. After the appropriate interval, replace the index-card shutter. Remove the film plate and take it to the teacher for developing. The room lights cannot be turned on until the film has been developed and bleached. Photographic glass plates should have a goldish color after developing. Overexposed plates appear bluish, and underexposed plates appear reddish. Remove your objects from the steel plate.

9. Repeat steps 6–8 for each lab group.

10. When your hologram is dry, hold it up to an overhead light or sunlight and look through it. **CAUTION:** *Never look directly into sunlight.* Record your observations in Table 1. Drying will be faster if you direct warm air from a hair dryer, slide projector, or overhead projector onto the plate.

11. Place a black background against the glass. Hold your hologram in front of a bright white light, such as that from a slide projector, an overhead projector, or sunlight. Look at the hologram and observe the reflected light. Record your observations in the table.

12. Wash your hands with soap and water before you leave the laboratory.

19-2 Physics Lab

Data and Observations

Table 1	
	Observations of Hologram
Looking through hologram	
Looking at reflection	

Analysis and Conclusions

1. Compare your observations of the hologram from the table. How do you account for the different observations?

_____

_____

_____

_____

2. Look at an ordinary photograph in your textbook or a magazine. How does the hologram compare to the photograph?

_____

_____

_____

3. The hologram was produced on black-and-white film. How was the color visible in the hologram produced?

_____

_____

_____

19-2 Physics Lab

Extension and Application

1. Use the vertical arrangement shown in Figure B to make a transmission hologram. Place the helping hand with the mirror near the laser, so that the laser beam reflects off it and the expanded beam shines onto the table. An index card serves again as the shutter. Use some small notebook-paper clips to hold the film perpendicular to the steel plate. This hologram must be viewed by looking through the hologram at the expanded laser beam. **CAUTION:** *Never look directly into the laser source or into an unexpanded beam.*

Figure B

A downward view of the arrangement of mirror, object, and holographic plate for production of a transmission hologram

Describe your transmission hologram and compare it to the white-light reflection hologram.

_____

_____

_____

_____

_____

2. Explain the value of using holograms in money, credit cards, or images of rare coins.

_____

_____

_____

_____

_____

20-1

Physics Lab

Investigating Static Electricity

Objectives

- **Demonstrate** that static charges can be separated by contact.
- **Demonstrate** that opposite charges attract and like charges repel.
- **Infer** the type of charge produced on various materials.
- **Compare** the type of charge produced on an object by induction and by conduction.

Clothes removed from a clothes dryer usually cling to each other and spark or crackle with static electricity when you separate them. When two dissimilar materials rub together, they can become charged. Objects acquire static-electric charges by either gaining or losing electrons. An object that gains electrons has a net negative charge. For example, a hard-rubber rod rubbed with wool becomes negatively charged. An object that loses electrons has a net positive charge. For example, a glass rod rubbed with silk becomes positively charged. Only those objects separated from a ground, or Earth, by an insulator will retain their charge for any length of time. Objects that are attached to a ground through a conductor stay uncharged, since the charge travels into the ground and quickly dissipates.

Procedure

Perform each part of the experiment several times to be sure you have proper observations of the phenomena. You may have to rub the rods vigorously to charge them, particularly the glass rods.

A. Negatively Charging a Pith Ball

1. Rub the hard-rubber rod with the wool pad to charge it. Bring the rod close to, but not touching, a suspended pith ball, as shown in Figure A. Observe the behavior of the pith ball and record your observations in Table 1. Touch the pith ball with your finger to remove any charge it may have.

2. Charge the rubber rod again. Bring it close to the pith ball and allow it to touch the ball. Then bring the charged rod near the charged ball and observe the behavior of the ball. Record your observations in Table 1.

Figure A

Materials

- hard-rubber rod
- glass rod
- plastic wrap
- silk pad
- wool pad
- pith ball suspended from holder by silk thread
- leaf or vane electroscope
- Leyden jar
- wire with alligator clips

B. Positively Charging a Pith Ball

1. Rub the glass rod with a piece of silk to charge it. Bring the rod close to, but not touching, a suspended pith ball. Observe the behavior of the pith ball and record your observations in Table 2. Touch the pith ball with your finger to remove any charge it may have.

20-1 Physics Lab

2. Charge the glass rod again. Bring it close to the pith ball and allow it to touch the ball. Then bring the charged rod near the charged ball and observe the behavior of the pith ball. Record your observations in Table 2.

C. Charging an Electroscope by Conduction

1. Charge the rubber rod with the wool. Gently bring it in contact with the metal top of the vane or leaf electroscope. The electroscope is now negatively charged. Observe what changes in the electroscope. Record your observations in Table 3.

2. Recharge the rubber rod. Bring it near the charged electroscope top. Observe what happens to the leaves or vane. Record your observations in Table 3.

3. Recharge the electroscope with a negative charge from the rubber rod. Charge the glass rod and bring it near the top of the negatively charged electroscope. Observe the electroscope deflection as you move the charged glass rod close to and away from the electroscope. Record your observations in Table 3. Discharge the electroscope by momentarily touching the top of it with your finger.

4. Charge the plastic wrap or acrylic rod with a piece of silk. Touch the plastic material to the top of the electroscope to charge it. Set the plastic aside. Charge the rubber rod and bring it near the electroscope while observing the deflection of the leaves or vane. Record your observations in Table 3.

D. Charging an Electroscope by Induction

1. Select one of the rods and charge it. Bring it within 1–2 cm of, but not touching, the electroscope. Observe the leaves or vane. Record your observations in Table 4.

Clip   Wire   Clip   **Figure B**

Vane electroscope   Leyden jar

2. With the charged rod near the electroscope, momentarily touch the top of the electroscope with your finger. Remove your finger from the top of the electroscope, and then remove the charged rod. Observe the electroscope leaves or vane. Record your observations in Table 4.

3. Following the procedure from Part C, determine the type of charge on the electroscope. Record your observations in Table 4.

20-1 Physics Lab

E. Charging a Leyden Jar

1. Place a Leyden jar beside the electroscope.

2. Charge one of the rods and gently touch the electroscope to charge it. Repeat the procedure to place a fairly large charge on the electroscope and to cause a large deflection.

3. Discharge the electroscope by momentarily touching the top with your finger. Use the clip wire to attach the top of the electroscope to the top of the Leyden jar, as shown in Figure B. Note that the Leyden jar is not grounded. Repeat step 2 to charge the combination. Is there any difference in the ability of the electroscope to become charged? Record your observations in Table 5.

Data and Observations

Table 1
Observations of pith ball with nearby charged rod
Observations of pith ball after it has been touched with a charged rod

Table 2
Observations of pith ball with nearby charged rod
Observations of pith ball after it has been touched with a charged rod

20-1 Physics Lab

Table 3
Observations of uncharged electroscope when negatively charged rod touches it
Observations of negatively charged electroscope when negatively charged rod is nearby
Observations of negatively charged electroscope when positively charged rod is nearby
Observations of deflection of leaves or vane when plastic material charges the electroscope and negatively charged rod is nearby

Table 4
Observations of electroscope when charged rod is nearby
Observations to electroscope after touching with finger
Observations to determine the type of charge

20-1 Physics Lab

Table 5
Observations of charging a Leyden–jar electroscope combination

Analysis and Conclusions

1. Summarize your observations of negatively charging a pith ball (Part A).

_____

_____

2. Summarize your observations of positively charging a pith ball (Part B).

_____

_____

3. Compare the negative and positive charging of a pith ball.

_____

_____

4. Summarize your observations of charging the electroscope by conduction (Part C).

_____

_____

_____

5. Using your data from Part C, explain why the leaves of the electroscope diverged or the vane deflected.

_____

_____

6. Using your data from Part C, explain why the leaves (vane) remained apart (deflected) when you removed the charged rod.

_____

_____

7. What type of charge did the plastic wrap/plastic rod produce? How can you prove this?

_____

_____

_____

8. Why did the electroscope vane or leaves move in Part D?

_____

_____

_____

_____

9. Compared to charge on the rod, what type of charge did the electroscope receive by induction?

_____

10. What purpose did your finger serve when it touched the electroscope?

_____

11. Compared to the charge of the charging body, what type of charge does an electroscope have when it is charged by conduction? By induction?

_____

_____

12. In Part E, how did the Leyden jar affect the ability to charge the electroscope? What is your proof of this effect?

_____

_____

_____

_____

Extension and Application

1. An electrostatic filter is sometimes used to filter the air in the heating system for a building. In this device, air passes between a number of thin metal plates or wires that are attached to a source of high voltage that generates a static charge on the plates. Dust and dirt particles are attracted to the plates or wires, thus leaving the air. Explain why the particles are attracted to the plates or wires, and what advantage this type of filter might have over an ordinary mechanical filter.

21-1

Physics Lab

The Capacitor

A capacitor stores electric charge. An early type of capacitor is the Leyden jar, which you used in Experiment 20-1. Capacitors are usually made of two conducting plates separated by some type of insulating material, called a dielectric. The dielectric can be any insulating material. Air, oil, glass, and ceramics are dielectric materials used in capacitors. The capacitance, or charge capacity, of a capacitor is directly proportional to the area of the plates and the material used for the dielectric, and inversely proportional to the distance separating the plates.

Figure A is a circuit diagram of a capacitor, a battery, a switch, a resistor, and an ammeter, connected in series. A CBL voltage probe is connected parallel to the battery. The resistor is a simple device that resists the flow of charge. The flow of electric charge over a period of time is measured in units called amperes or amps; 1 coulomb/second = 1 ampere. When the switch is open, as shown in figure A, no charge flows from the battery. When the switch is closed, the battery supplies electric energy to move positive charges to one plate of the capacitor and negative charges to the other. Charge accumulates on each plate of the capacitor, but no current flows through it because its center is an insulator. As charge accumulates in the capacitor, the potential difference increases between the two plates until it reaches the same potential difference as that of the battery. At this point, the system is in equilibrium, and no more charge flows to the capacitor.

Objectives

- **Demonstrate** the storage of electric charge in a capacitor.
- **Observe** the characteristic curve of the discharge of voltage from a capacitor over time.
- **Infer** that electric current does not flow through a capacitor.
- **Predict** the charging time for a capacitor without series resistance.

Materials

- 1000-μF capacitor
- 10-kΩ resistor
- 27-kΩ resistor
- 15-V DC source
- SPDT knife switch
- connecting wire
- CBL unit
- link cable
- graphing calculator
- voltage probe

Capacitance is measured by placing a specific amount of charge on a capacitor and then measuring the resulting potential difference. The capacitance, C, is given by

$$C = q/V,$$

where C is the capacitance in farads, q is the charge in coulombs, and V is the potential difference in volts.

Figure A

Procedure

1. Set up the circuit shown in Figure A. The ammeter, capacitor, and battery must be wired in the proper order. Look for the + and − marks on the circuit components. The positive plate of the capacitor must be wired to the positive terminal of the battery. If the connections are reversed, the capacitor may be destroyed. Resistors have no + or − end. Record the capacitance.

21-1 Physics Lab

2. Connect the voltage probe to port CH1 of the CBL unit. Attach the link cable from the CBL unit to the graphing calculator. Start the program PHYSICS in the graphing calculator. From the MAIN MENU, select SET UP PROBES. Enter 1 as the number of probes. From the SELECT PROBE menu that appears, select MORE PROBES and then choose VOLTAGE. Next, enter the CHANNEL NUMBER as 1.

3. From the MAIN MENU, select COLLECT DATA. Next, select TIME GRAPH from the DATA COLLECTION menu. For the time between samples in seconds, enter 1. For the number of samples, enter 99. Next, select USE TIME SETUP, and then chose NON-LIVE DISPLAY.

4. Charge the capacitor to the battery voltage by closing the knife switch to position 1. Let the capacitor charge for two to three minutes before proceeding.

5. Press ENTER on the graphing calculator to begin collecting data. Immediately move the knife switch to position 2 to bypass the battery and begin the capacitor discharging.

6. When the CBL unit indicates DONE, press ENTER. A display of voltage verses time will appear. Use the right arrow key to move through the data, with X being the time and Y being the voltage value. Record the corresponding voltage readings with the times in Table 1.

7. Replace the 27-kΩ resistor with the 10-kΩ resistor.

8. Repeat steps 2–6 with the 10-kΩ resistor.

Data and Observations

Capacitance: _____

Table 1					
	27 kΩ	10 kΩ		27 kΩ	10 kΩ
Time (s)	Voltage (V)	Voltage (V)	Time (s)	Voltage (V)	Voltage (V)
0			50		
5			55		
10			60		
15			65		
20			70		
25			75		
30			80		
35			85		
40			90		
45					

21-1 Physics Lab

Analysis and Conclusions

1. Why did the voltage start at a maximum value and drop toward zero while the capacitor was discharging?

   _____

   _____

   _____

2. Look at the data for the two resistors. Explain the purpose of the resistor in the circuit.

   _____

   _____

   _____

3. Using the data in Table 1, plot two graphs for voltage as a function of time. Use a separate sheet of graph paper. Sketch a smooth connecting curve.

4. By Ohm's law, $I = \frac{V}{R}$. Therefore the charge, Q, stored in the capacitor, is given by $Q = It = (\frac{V}{R})t$, which is the area between the curve and the time axis. Estimate the area under the curve by sketching one or more triangles to approximate the area. What is the estimated charge for the capacitor with the 27-kΩ resistor and with the 10-kΩ resistor?

5. Calculate the capacitance of the capacitor, $C = q/V$, using the value for charge from Question 4 and the measured potential difference of the power source.

6. Compare the value determined in Question 5 with the manufacturer's stated value you recorded in Data and Observations. Electrolytic capacitors have large tolerances, frequently on the order of 50%, so there may be a sizable difference. Find the relative error between the two values.

7. Predict what would happen if the circuit were set up without the resistor.

   _____

   _____

21-1 Physics Lab

8. Which variables were manipulated and which remained constant during the experiment? Describe the voltage versus time curve.

_____

_____

Extension and Application

1. Repeat the experiment with another capacitor obtained from your teacher, or connect several capacitors in parallel. Capacitors connected in parallel have an effective value equal to the sum of the individual capacitors. The quantity *RC* is the time constant for the circuit and has units of seconds, as long as *R* is measured in ohms and *C* is measured in farads. The time constant is the time required for the current to drop to 37% of its original value. Before you begin, check to see that the value of *RC* can be easily measured. Follow the same procedure as before, and compare the results of the graph with your earlier results.

2. Describe how an RC circuit (a circuit that includes a resistor and a capacitor), capable of charging and discharging at a very specific and constant rate, could be in the home, an automobile, or an entertainment center.

_____

_____

_____

22-1

Physics Lab

Ohm's Law

Ohm's law states that, as long as the temperature of a resistor does not change, the electric current, I, in a circuit is directly proportional to the applied voltage, V, and inversely proportional to the resistance R:

$$I = \frac{V}{R}; \text{ thus } R = \frac{V}{I}$$

The resistance is measured in ohms (Ω), where $1\Omega = 1$ V/1 A.

Procedure

1. Study Figure A. Notice in the figure the arrangement of the probes, switch, resistor, and power supply. Compare the arrangement with the schematic diagram of the same circuit in Figure B. Select a 100-Ω resistor. Refer to the Resistor Color Code on page xviii to determine the values for the various color bands. Determine the tolerance range for the resistor and record this value in the table. Connect the circuit, knife switch open, as shown in Figure A; use the 100-Ω resistor. Leave the knife switch open so that no current flows in the circuit until your teacher has checked your circuit and has given you permission to proceed.

2. Connect the CBL unit and graphing calculator with the unit-to-unit link cable. Connect a CBL-DIN adapter to CH1 and CH2 ports on the CBL unit. On the amplifier box, connect the DIN1 lead to the CBL port CH1 and the DIN2 lead to port CH2. Connect the current1 probe to PROBE1 port on the amplifier box, and connect the voltage1 probe to PROBE2 port of the amplifier box. Turn on the CBL unit and your graphing calculator. Start the PHYSICS program.

3. From the MAIN MENU, select SET UP PROBES. Enter 2 as the number of probes. Select MORE PROBES from the SELECT PROBE menu. Select the C-V CURRENT probe from the list. Enter 1 as the channel number. Select USE STORED from the calibration menu. Again from the SELECT PROBE menu, choose MORE PROBES and then select C-V VOLTAGE. Enter channel number 2 for the channel number. Select USE STORED from the CALIBRATION menu.

4. Select the COLLECT DATA option from the MAIN MENU. From the DATA COLLECTION menu, select MONITOR INPUT. The calculator will begin displaying the inputs from both channels.

5. To prevent the resistors from overheating, the knife switch should only be closed long enough to obtain readings of current and voltage. Close the switch and adjust the voltage to about 1.5 V. If you are using a battery, there is no adjustment. Open the knife switch.

Objectives

- **Measure** the current flowing through a resistor and the voltage across a resistor.
- **Calculate** resistance from the measured values.
- **Analyze** the relationship between voltage and current for a resistor.

Materials

- power supply or battery: two- 1.5 V cells
- connecting wire
- knife switch
- resistors
- CBL unit
- link cable
- voltage and current probe with dual channel amplifier box
- 2 CBL-DIN adapters
- graphing calculator

22-1 Physics Lab

Figure A

Be sure to connect the voltmeter probe in parallel with the resistor and the ammeter in series with the resistor.

Figure B

The circuit diagram for the apparatus is shown in Figure B.

Record your readings in Table 1. Close the switch and adjust the voltage to a higher reading, such as 3.0 volts. Do not exceed 5.0 V. If you are using a second battery, add it with the knife switch open. Open the knife switch and record your reading.

6. Remove the 100-Ω resistor and replace it with the 150-Ω resistor. Repeat step 5 to obtain two sets of data for the current and voltage.

7. Remove the 150-Ω resistor and replace it with the 220-Ω resistor. Repeat step 5 to obtain two sets of data for the current and voltage.

Data and Observations

Table 1						
Resistor	Printed value of resistor (Ω)	Tolerance range (+/− %)	Voltage (V)	Current (mA)	Current (A)	Resistance (Ω)
R_1						
R_2						
R_3						

22-1 Physics Lab

Analysis and Conclusions

1. Use your data to calculate the resistance for each set of data, applying $R = V/I$ where V is in volts and I is in amperes. Record the resistances in the table.

2. Calculate the average of two values for each resistor. Compare the printed values of the resistors you used with your averages. Determine the relative error for each resistor.

3. If your values were not within the tolerance range of the printed values, suggest some reasons for the discrepancy.

_____

_____

_____

_____

4. Describe the proper placement of an ammeter in a circuit.

_____

_____

_____

5. Describe the proper placement of a voltmeter in a circuit.

_____

_____

_____

6. State the relationship between the current through a circuit and the voltage and resistance of the circuit.

_____

_____

_____

22-1 Physics Lab

Extension and Application

1. From your teacher, obtain a resistor of unknown value. Connect the circuit components in proper order and take several voltage and current readings through this resistor. Plot a graph of voltage (*y*-axis) versus current (*x*-axis). Sketch a smooth line that best fits all the data. Determine the resistance and calculate the slope of the line, using volts for potential difference and amperes for current.

2. A lightbulb, after it has reached its operating temperature, has a potential difference of 120 V applied across it, while 0.5 A of current passes through it. What is the resistance of the lightbulb? When the lightbulb is cold, it has a resistance of 21 Ω. What is the current, in amperes, at the instant the bulb is connected to a potential difference of 120 V?

22-2

Physics Lab

Electric Equivalent of Heat

Objectives

- **Demonstrate** the law of conservation of energy.
- **Measure** and **compare** quantities of electric energy and heat energy.
- **Calculate** the experimental error.

The law of conservation of energy states that energy changes from one form to another without loss or gain. This means that 1J of potential energy, when converted to electricity, should develop 1J of electric energy. 1J of electric energy, when converted to thermal energy, should produce 1J of thermal energy. Electric power is $P = IV$, where I is the current in amperes and V is the potential difference in volts. Electric energy is power multiplied by time, so $E = Pt = IVt$, where E is the energy in joules and t is the time in seconds. The thermal energy in water can be written as $Q_w = m_w C_w \Delta T_w$, where Q is the thermal energy in joules, m_w is the mass of the water, C_w is the specific heat of water, and ΔT_w is the temperature change of the water.

In this experiment, you will measure the amount of electric energy converted into thermal energy by means of an electric heating coil (a resistor) immersed in water. You will also measure the amount of heat absorbed by a known mass of water. The polystyrene cup, used as a calorimeter, has a negligible specific heat and will not absorb enough thermal energy to affect the experiment.

To minimize the effect of heat loss to the atmosphere, warm the water to as many degrees above room temperature as it was below room temperature before heating started. Thus, if you start with water at 10°C and room temperature is 20°C, the final temperature of the water should be 30°C. In this way, any heat gained from the surroundings while the temperatures are below room temperature is likely to be offset by an equal heat loss while temperatures are higher than room temperature.

Materials

- large polystyrene cup
- power supply
- connecting wire
- DC ammeter
- rheostat
- voltmeter
- Celsius thermometer
- balance
- stopwatch

Procedure

1. Measure the mass of a polystyrene cup and record this value in Table 1. Record the room temperature.

2. Fill the cup about two-thirds full of cold water. Measure the mass of the cup plus water. Record this value in Table 1. Calculate and record the mass of the water.

3. Set up the circuit, as shown in Figure A. If you have an adjustable power supply, the rheostat is built in, rather than separate. If the coil is not immersed, add more water and repeat step 2. After the teacher has checked your circuit, close the switch. Adjust the rheostat until the current flow is 2–3 A. Open the switch at once.

4. Stir the water gently with the thermometer. Read the initial temperature of the water. Record this value in Table 1.

22-2 Physics Lab

5. Prepare to time your readings. Close the switch. At the end of each minute, read the ammeter and voltmeter readings and record these values in Table 2. Gently stir the water from time to time and, if necessary, adjust the rheostat to maintain a constant current flow.

6. Monitor the water temperature to determine when it is approximately as many degrees above room temperature as it was below room temperature at the beginning of the experiment. At the end of the next minute after this temperature is reached, open the switch.

7. Stir the water gently until it reaches a constant temperature. Record in Table 1 the final temperature of the water. Compute and record the change in temperature of the water.

8. Determine the average current and the average voltage. Record these values in Table 1.

Figure A
(a) The heating coil changes electric energy to heat energy that is transferred to the water.
(b) Circuit diagram of the apparatus.

Data and Observations

Table 1	
Mass of polystyrene cup	
Mass of water and cup	
Mass of cold water	
Initial temperature of water	
Room temperature	
Final temperature of water	
Change in temperature of water	
Average current	
Average voltage	

22-2 Physics Lab

Table 2					
Time (min)	Current (A)	Voltage (V)	Time (min)	Current (A)	Voltage (V)
1			9		
2			10		
3			11		
4			12		
5			13		
6			14		
7			15		
8					

Analysis and Conclusions

1. Determine the electric energy expended in the resistance, using $E = IVt$.

2. Determine the heat absorbed by the water, using $Q_w = m_w C_w \Delta T_w$, where C_w is 4.18 J/g °C.

3. Find the relative difference between the electric energy expended and the thermal energy absorbed by the water. Use % difference = $(E - Q_w)(100\%)/E$.

4. Taking into account the apparatus you used (especially the heating coil), suggest why the difference in question 3 is not zero.

_____

_____

_____

22-2 Physics Lab

• • • • • • • • • • • • • • •

5. Were your results close enough to indicate that, under ideal conditions, you would find exact agreement in the energy exchange? Give a reason for your answer.

_____

_____

_____

6. What percentage of the electric energy was converted to thermal energy in the water?

Extension and Application

1. Think about using the conversion of gravitational potential energy to thermal energy in order to heat water for showers. Assume that 1.0 L of water (1.0 kg of water), initially at 12.7°C, falls over a 50.0-m waterfall. If all of the potential energy of the water is converted to kinetic energy as the water is falling, and the kinetic energy is converted into thermal energy when it reaches the bottom, what temperature difference exists between the bottom and the top of the waterfall? Will this be sufficient to produce a warm shower? Show your calculations.

2. An electric immersion heating coil is used to bring 180.0 mL of water to a boil for a cup of tea. If the immersion heater is rated at 200 W, find the time needed to bring this amount of water, initially at 21°C, to the boiling point. Show your calculations.

3. Of the electric energy used by a 100-W incandescent lightbulb, 16 percent is converted to light. How many joules of heat energy dissipate each second?

23-1

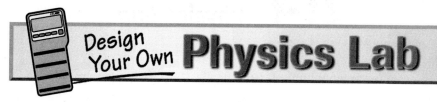

Design Your Own **Physics Lab**

Series Resistance

Objectives

- **Measure** the voltage across each resistor in a series circuit.
- **Measure** the current in a series circuit.
- **Observe** the voltage drop across a resistor and the total voltage drop across the series circuit.
- **Determine** the relationship between the voltage drop across a resistor and its resistance.

Possible Materials

- power supply with variable voltage
- wires with clips
- resistors, 100 Ω, 150 Ω, 220 Ω, 470 Ω
- voltmeter
- ammeter
- knife switch
- CBL unit
- link cable
- voltage and current probe with dual-channel amplifier box
- 2 CBL-DIN adapters
- graphing calculator

A series circuit has two or more devices connected so that all the current in the circuit flows through each device in turn. When all the devices are resistors, you have a series resistance circuit. In a properly constructed series circuit, a device can't have a different amount of current through it than any other device in the circuit. Think of a series circuit as if it were several pieces of hose with different diameters, connected end to end. The volume of water that flows out of the hose must be the same as the volume that enters the hose, no matter what the diameter. Each resistance in a series circuit can be analyzed using the relationship $I = V/R$. And the voltage across the entire series circuit must be equal to the sum of the voltages across each part of the circuit.

Series resistance circuits have many electric and electronic applications. They are most often used to divide a voltage into two or more smaller voltages, such as allowing one battery or power supply to provide more than one value of voltage. Series circuits can also reduce a large voltage to a smaller one, which is how it is possible for a multimeter to measure a wide range of voltages.

Problem

What is the relationship between the voltage drops across resistors in a series circuit and the total voltage across the circuit?

Hypothesis

Formulate a hypothesis about the relationship between voltage across resistors in a circuit and the circuit's total resistance.

_____

_____

Plan the Experiment

1. Decide on a procedure that uses the suggested materials (or others of your choosing) to measure voltages in a series resistance circuit that includes a source of voltage. Using the relationship $I = V/R$, predict the relationship between the current through the circuit and the voltage across each resistor in the circuit.

2. On the next page, draw a series circuit with three resistors that you will use for the experiment. Indicate positions where you might make voltage or current measurements. Remember that voltmeters must be wired in parallel and that ammeters must be wired in series.

23-1 Physics Lab

3. Decide what kind of data to collect and how to analyze it. You can record your data in the table below. Label the columns appropriately.

4. Write your procedure on another sheet of paper or in your notebook.

5. **Check the Plan** Have your teacher approve your plan before you proceed with your experiment.

Series Circuit with Three Resistors

Data and Observations

Data Table							

23-1 Physics Lab

Analyze and Conclude

1. **Analyzing Results** What is the relationship between the voltage drop across the individual resistors in the circuit and the power-supply voltage?

   _____

   _____

   _____

2. **Analyzing Results** Is the voltage drop across each resistor related to the value of the resistor? Give a reason for your answer.

   _____

   _____

   _____

   _____

3. **Checking Your Hypothesis** Would removing a resistor affect the voltage drops across the remaining resistors in a series resistance circuit? Give a reason for your answer.

   _____

   _____

   _____

   _____

4. **Checking Your Hypothesis** Describe the total current in a series resistance circuit in relation to the total resistance and to the voltage applied to the circuit.

   _____

   _____

   _____

   _____

5. **Predicting** Predict the equivalent resistance of a circuit when the circuit consists of resistors in series. Use several resistors and an ohmmeter to check your prediction.

   _____

   _____

   _____

   _____

23-1 Physics Lab

Apply

1. A set of 100 lightbulbs wired in series lines a sidewalk. The set of lights is designed to operate on 120 V. If the set uses 0.5 A of current, what is the average resistance of each individual bulb? What is the average voltage across each lightbulb?

2. You are asked to determine the resistance value of an unmarked resistor. You have a voltmeter, a battery, and several resistors of known value. Explain the method you would use to determine the value of the unmarked resistor using only these items.

_____

_____

_____

_____

_____

_____

_____

_____

23-2

Physics Lab

Parallel Resistance

Objectives

- **Compare** the total current in a parallel circuit to the current in each resistor.

- **Infer** the difference in total resistance of a circuit after adding resistors.

- **Demonstrate** changes in total current in a circuit after adding resistors in parallel.

- **Calculate** the equivalent resistance of a parallel resistance circuit.

When resistors are connected in parallel, each resistor provides a path for current and so reduces the equivalent resistance. In parallel circuits, each element has the same applied potential difference. In Figure A3, three resistors are connected in parallel across the voltage source. The current can pass from junction a to junction b along three paths. More current will flow between these junctions than would flow if only one or two resistors connected them. The total current, I, is represented by the following equation.

$$I = I_1 + I_2 + I_3$$

Each time another resistance is connected in parallel with other resistors, the equivalent resistance decreases. The equivalent resistance of resistors in parallel can be determined by the following equation.

$$\frac{1}{R} = \frac{1}{R_1} + \frac{1}{R_2} + \frac{1}{R_3} + \ldots$$

In this experiment, you will measure current and voltage with resistors in parallel and apply the relationship $R = V/I$ to verify your results. Follow closely the circuit diagrams in Figure A, but because you probably have only one ammeter and one voltmeter, you must move the meters from position to position to get your readings. For example, take the total current and total voltage readings, then move the meters to positions I_1 and V_1, and so on. For your calculations, convert milliampere readings to amperes (1 mA = 0.001 A).

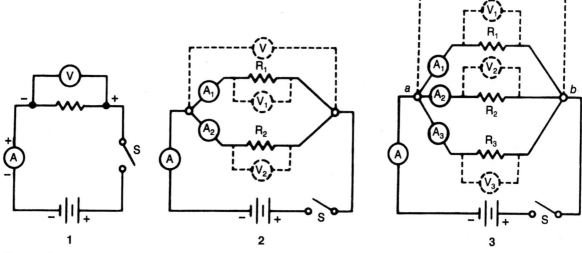

Figure A

Resistors are connected in parallel in 2 and 3. Note the positive and negative terminals of the voltmeter and ammeter in relation to the positive and negative terminals of the voltage source.

23-2 Physics Lab

Materials

DC power supply or dry cells

3 resistors, 0.5-W, 150–330-Ω range, such as 180 Ω, 220 Ω, and 330 Ω

connecting wires

knife switch

voltmeter, 0–5 V

milliammeter, 0–50 mA or 0–100 mA

Procedure

A. One Resistor

1. Set up the circuit as shown in Figure A1, using one resistor. Close the switch. Adjust the power supply to a set voltage on the voltmeter, such as 3.0 V. Read the current value on the ammeter. Open the switch. Record your readings and the resistor tolerance in Table 1.

B. Two Resistors

1. Set up the circuit as shown in Figure A2. Close the switch and adjust the power supply to maintain the same voltage reading as in Part A. Read the current value on the ammeter. Open the switch. Record your readings in Table 2.

2. Move the meters to get the other readings. Record the readings and tolerances in Table 2.

C. Three Resistors

1. Set up the circuit as shown in Figure A3. Close the switch, adjust the power supply, and read the meters. Open the switch. Record the readings in Table 3.

2. Move the meters to get the other readings. Record the readings and tolerances in Table 3.

Data and Observations

Table 1			
	R_1 (Ω)	Ammeter reading (mA)	Voltmeter reading (V)
Tolerance (%)			

Table 2							
	R_1 (Ω)	R_2 (Ω)	Ammeter reading (mA)			Voltmeter reading (V)		
			I	I_1	I_2	V	V_1	V_2
Tolerance (%)								

23-2 Physics Lab

Table 3											
	R_1 (Ω)	R_2 (Ω)	R_3 (Ω)	Ammeter reading (mA)				Voltmeter reading (V)			
				I	I_1	I_2	I_3	V	V_1	V_2	V_3
Tolerance (%)											

Analysis and Conclusions

1. Use data from Table 1 to calculate R_1, where $R_1 = \dfrac{V}{I}$. Is this within the tolerance expected for the printed value for R_1?

2. Use data from Table 2 to calculate the following values:

   a. the measured equivalent resistance, R, where $R = \dfrac{V}{I}$.

   b. the current, $I_1 + I_2$.

   c. the resistance of resistor 1, where $R_1 = \dfrac{V_1}{I_1}$.

   d. the resistance of resistor 2, where $R_2 = \dfrac{V_2}{I_2}$.

   e. the equivalent resistance, R, where $\dfrac{1}{R} = \dfrac{1}{R_1} + \dfrac{1}{R_2}$.

23-2 Physics Lab

• • • • • • • • • • • • • • •

3. a. Compare the current sum, $I_1 + I_2$, to the measured current, I.

_____

_____

b. Compare the calculated equivalent resistance to the measured equivalent resistance. Is the measured equivalent resistance within the tolerance range for the resistors?

_____

_____

4. Use data from Table 3 to calculate the following:

a. the equivalent resistance, R, where $R = \dfrac{V}{I}$.

b. the current, $I = I_1 + I_2 + I_3$.

c. the resistance of resistor 1, where $R_1 = \dfrac{V_1}{I_1}$.

d. the resistance of resistor 2, where $R_2 = \dfrac{V_2}{I_2}$.

e. the resistance of resistor 3, where $R_3 = \dfrac{V_3}{I_3}$.

f. the equivalent resistance, R, where $\dfrac{1}{R} = \dfrac{1}{R_1} + \dfrac{1}{R_2} + \dfrac{1}{R_3}$.

5. a. Compare the value of I to the measured current sum, $I_1 + I_2 + I_3$.

_____

_____

  b. Compare the calculated equivalent resistance to the measured equivalent resistance. Is the measured equivalent resistance within the tolerance range for the resistors?

_____

_____

6. How does the current in the branches of a parallel circuit relate to the total current in the circuit?

_____

_____

_____

7. How does the voltage drop across each branch of a parallel circuit relate to the voltage drop across the entire circuit?

_____

_____

_____

8. As more resistors are added in parallel to an existing circuit, what happens to the total circuit current?

_____

_____

Extension and Application

1. Tom has a sensitive ammeter that needs only 1.000 mA to provide a full-scale reading. The resistance of the ammeter coil is 500.0 Ω. He wants to use the meter for a physics experiment that needs an ammeter that can read 1.000 A. He has calculated that an equivalent resistance of 0.5000 Ω will produce the necessary voltage drop of 0.5000 V ($V = IR = 1.000 \times 10^{-3}$ A $\times$ 500.0 Ω), so that only 1.000 mA of current passes through the meter. What value of shunt resistor, a resistor placed in parallel with the meter, should he use?

23-2 Physics Lab

2. A voltmeter has resistance and provides a path for current in the circuit being measured. It is often important to know the resistance of the voltmeter, especially when measuring the voltage across a resistor with very little current or a resistor with high resistance. Assume that the current in a circuit is constant and that you want to measure the voltage across a 1000-Ω resistor. Would a voltmeter with a resistance of 10 000 Ω be a good choice? What about a voltmeter with a resistance of 1 000 000 Ω? Give reasons for your answer.

_____

_____

_____

_____

_____

_____

_____

3. A unique cube of 100-Ω resistors, shown in Figure B, is attached to a 1.0-V battery. The circuit can be reduced to combinations of parallel and series resistances. Predict the equivalent resistance. What is the total current supplied by the battery? Using twelve 100-Ω resistors, construct the circuit, and test your prediction.

1.0 V

Figure B

24-1

The Nature of Magnetism

Objectives

- **Determine** the polarity of a magnet.
- **Observe** the magnetic fields around magnets.
- **Experiment** with induced magnetism.

Many substances have magnetic properties. For example, liquid oxygen can be attracted by a magnet. However, only iron, nickel, and cobalt, and alloys that contain them, make strong permanent magnets. A magnet close to an object that contains one of these metals will induce magnetism in the object. This induced magnetism shows up as an attraction between the magnet and the object; that's why a magnet attracts a plain iron nail.

The forces between a magnet and magnetic objects or other magnets can be explained in terms of a magnetic field. Iron filings provide a convenient way to show the magnetic field pattern around a magnet. Magnetic induction makes the filings temporary magnets that align themselves with the field of the magnet that caused the induction. The long, thin filings act like a compass needle, pointing along the magnetic field lines.

A compass is a small magnet that is free to pivot in a horizontal plane about an axis. The end of the magnet that points to geographic north is called the north (N) pole. The opposite end of the magnet is the south (S) pole. Stated another way, a magnet's N pole is really the geographic north-seeking pole. By definition, magnetic field lines exit the N pole of a magnet and enter the S pole.

Problem

How can you determine the polarity of a magnet and map its magnetic field?

Hypothesis

Formulate a hypothesis about using a compass to find the polarity of a magnet and to map magnetic field lines.

Possible Materials

- 2 bar magnets
- sheet of paper or thin cardboard
- magnetic compass
- iron filings
- iron nail

Plan the Experiment

1. Decide on a procedure that uses the suggested materials (or others of your choosing) to determine the polarity of a bar magnet and the shape and direction of the field about the magnet. Make sure you investigate fields around pairs of like and unlike poles.

2. Decide what kind of data to collect and how to analyze it. You can record your data on the drawings on the next page.

3. Write your procedure on another sheet of paper or in your notebook.

4. **Check the Plan** Have your teacher approve your plan before you proceed with your experiment.

24-1 Physics Lab

Data and Observations

Magnetic field lines between poles

N		N

N		S

Direction and shape of magnetic field lines

N		S

Analyze and Conclude

1. **Analyzing Results** Where are the magnetic field lines around a magnet most concentrated?
   Describe the patterns formed by the magnetic field lines between like poles and unlike poles.

   _____

   _____

   _____

   _____

24-1 Physics Lab

• • • • • • • • • • • • • •

2. **Checking Your Hypothesis** How does the orientation of a compass needle change relative to the magnetic field of a bar magnet?

_____

_____

_____

3. **Checking Your Hypothesis** When a compass points to the south pole of a magnet, what do the magnetic field lines around the compass and magnet look like?

_____

_____

_____

4. **Predicting** A magnet is attached to one end of an iron nail. Predict the polarity of the induced magnetism at the free end of the nail compared to the magnet. Use a compass to test your prediction.

_____

_____

_____

Apply

1. Draw a figure that represents Earth. Label the north and south Poles. The interior of Earth can be thought of as a bar magnet with one pole at the magnetic north Pole and one pole at the magnetic south Pole. Add to your sketch a bar magnet showing the correct polarity for Earth, and sketch the magnetic field lines around Earth.

24-1 Physics Lab

2. Describe how you would determine what the magnetic field of a refrigerator magnet looks like. Try your method, and report the results.

_____

_____

_____

_____

_____

_____

_____

_____

24-2

Physics Lab

Principles of Electromagnetism

While experimenting with electric currents in wires, Hans Christian Oersted discovered that a nearby compass needle is deflected when current passes through the wires. This deflection indicates a magnetic field around the wire. You have learned that a compass shows the direction of the magnetic field lines. Likewise, the first right-hand rule determines the direction of the magnetic field: when the right thumb points in the direction of the conventional current, the fingers point in the direction of the magnetic field.

When an electric current flows through a loop of wire, a magnetic field appears around the loop. An electromagnet can be made by winding a current-carrying wire around a soft iron core. A wire looped several times forms a coil. The coil has a field like that of a permanent magnet. A coil of wire wrapped around a core is a solenoid. The iron core collects the magnetic field lines around the wire windings, creating a strong magnet.

Procedure

A. The Field Around a Long, Straight Wire

1. Place the cardboard at the edge of the lab table. Arrange the wire so that it passes vertically through a hole in the center of the cardboard, as shown in Figure A. Position the ring stand and clamp so that the wire continues vertically from the hole to the clamp. Bring the wire down the clamp and ring stand to the ammeter, and then to the positive terminal of the power supply.

The wire should pass vertically at least 10 cm below the cardboard before running along the table to the knife switch and then to the negative terminal of the power supply. Observe the proper polarity of the power supply and the ammeter as you connect the wires.

Objectives

- **Observe** the magnetic field around a wire that carries current.
- **Demonstrate** the relationship between current and magnetic field strength.
- **Observe** the relationship between magnetic field polarity and direction of current flow.
- **Relate** the magnetic field strength of a iron-core coil to that of an air-core coil.

Materials

- compass
- DC power supply
- ammeter
- 14–18 gauge wire
- large iron nail
- knife switch
- ring stand
- ring stand clamp
- iron filings
- cardboard sheet
- paper
- meterstick
- masking tape
- box of steel paper clips

Figure A

24-2 Physics Lab

2. Close the switch and set the current to 2–3 A. Open the switch. Place the compass next to the wire. **CAUTION:** *The wire may become hot if current flows for very long Close the switch only long enough to make your observations.* Move the compass about the wire to map the field. In Part A of Data and Observations, sketch the field around the wire.

3. Reverse the connections at the power supply and ammeter so that the current will flow in the opposite direction. Close the switch and use the compass to map the field around the wire. Sketch the field around the wire.

B. Strength of the Field

1. Make a slit and a hole in a piece of paper and put it over the cardboard with the wire at the center. Sprinkle some iron filings on the paper around the wire.

2. Close the switch and set the current to about 4 A. Gently tap the cardboard several times with your finger. Turn off the current. In Part B of Data and Observations, record your observations.

3. Tap the cardboard to disarrange the filings. Turn on the current and reduce it to 2.5 A. Gently tap the cardboard several times with your finger. Record your observations.

4. Tap the cardboard to disarrange the filings. Turn on the current and reduce it to 0.5 A. Gently tap the cardboard several times with your finger. Record your observations. Return the iron filings to the container.

C. The Field Around a Coil

1. Remove the straight wire from the cardboard. In the center of the wire, make three loops of wire around your hand to form a coil of wire with a diameter of approximately 10 cm. Tape the loops together in a few spots.

2. Connect the coil to the power supply through the ammeter and knife switch. Close the switch and adjust the current to 2.5 A. Hold the wire coil in a vertical plane and bring the compass near the coil, moving it through and around the coil of wire. In Part C of Data and Observations, sketch the magnetic field direction around the coil. Show the positive and negative connections of your coil.

D. An Electromagnet

1. Uncoil your large loops. Wind loops of wire around the nail or another iron core until about half of the core is covered. Connect the coil to the power supply, knife switch, and ammeter. Adjust the current so that 1.0 A of current flows in the coil. Bring the core near the box of paper clips and see how many paper clips the electromagnet can pick up. Open the switch and record your observations in Part D of Data and Observations.

2. Wind several more turns of wire on the iron core to double the number of turns from step 1. Close the switch and see how many paper clips the electromagnet can pick up. Open the switch and record your observations.

3. Increase the current to 2.0 A and see how many paper clips the electromagnet can pick up. Record your observations.

4. Turn the current on and, using the compass, determine the polarity of the electromagnet.

Copyright © by Glencoe/McGraw-Hill

Physics: Principles and Problems

24-2 Physics Lab

Data and Observations

A. The Field Around a Long, Straight Wire
Observations of direction of north pole:

B. Strength of the Field
Observations of magnetic field with different currents:

C. The Field Around a Coil
Observations of magnetic field around a coil of current-carrying wire:

D. An Electromagnet
1. Several loops of wire on iron core:

_____

2. Double the number of loops:

_____

3. Double the current:

_____

Analysis and Conclusions

1. How does the right-hand rule apply to the current in a long, straight wire?

_____

_____

_____

_____

2. What is the effect of increasing the current in a wire?

_____

3. In Figure B draw the magnetic field direction and the polarities for the current-carrying coil.

Figure B

4. What three factors determine the strength of an electromagnet?

_____

_____

5. Explain the difference between a bar magnet and an electromagnet.

_____

_____

_____

Extension and Application

1. List several solenoid applications that operate with either continuously applied or intermittently applied currents.

_____

_____

_____

_____

24-3

Design Your Own Physics Lab

Objectives

- **Compare** the strength of the magnetic force exerted by electromagnets having different numbers of wire turns.

- **Predict** the relationship between the number of wire turns in an electromagnet and the strength of the magnetic force.

- **Calculate** the constant of variation for the relationship of wire turns to magnetic force.

Possible Materials

- insulated magnet wire
- iron bolts
- paper cups
- ammeter
- DC power supply
- iron BBs
- masking tape
- connecting wires
- marking pen

Variation in the Strength of Electromagnets

Any conductor that carries an electric current has a magnetic field around it, and this field exerts a magnetic force. When the conductor forms a coil, the magnetic field exerts more force on magnetic materials. A coil of wire with a core of magnetic material is an electromagnet.

Several factors affect the strength of an electromagnet. A magnet with a core of a special metal alloy containing iron and other elements has a far stronger magnetic field than a core of iron alone. The number of loops in the coil affects the strength of the magnetic field. Another factor that affects the strength of an electromagnet is the way the coil is wound. A coil with turns that are spread out over a core usually generates a weaker magnetic field than the same number of turns wound close together on the same core.

The strength of a magnetic field is measured in teslas. One tesla equals a force of 1 N on a straight conductor 1 m long, carrying a current of 1 A. As a practical matter, you can measure the force exerted by a magnetic field by determining how much mass the magnetic field can hold against the force of gravity. With this method, you can determine the relationship between the number of wire turns in a coil and the force of the resulting magnetic field. In this lab, you will determine the relationship between the number of wire turns in an electromagnet and the force exerted by the magnetic field of the electromagnet.

Problem

What is the relationship between the strength of the magnetic field of an electromagnet and the number of wire turns in the coil of the electromagnet?

Hypothesis

Formulate a hypothesis about the strength of a magnetic field in an electromagnet that will allow you to predict how large a weight you can hold with the magnet.

Plan the Experiment

1. Decide on a procedure that uses the suggested materials (or others of your choosing) to obtain data on how the force exerted by the electromagnet relates to the force of gravity.

2. Decide what kind of data to collect and how to analyze it. You can record your data in the table on the next page. Label the columns appropriately.

3. Write your procedure on another sheet of paper or in your notebook. On the next page, draw the setup you plan to use.

4. **Check the Plan** Have your teacher approve your plan before you proceed with your experiment.

Setup

Data and Observations

Data Table			

24-3 Physics Lab

Analyze and Conclude

1. **Analyzing Results** How is the strength of the magnetic force of an electromagnet related to the number of wire turns on the magnet?

   _____

   _____

2. **Formulating Models** If x is the number of BBs, y is the number of wire turns, and k is a constant of variation, write an equation that shows the relationship between the number of BBs and the number of wire turns.

3. **Using Numbers** Use a calculator and the equation from question 2 to find values of k for each of your electromagnets. Record these in your data table. What can you say about the calculated values?

   _____

   _____

   _____

4. **Inferring** What factor(s) in your setup should be kept constant to give the most consistent and reliable results? Give a reason for your answer.

   _____

   _____

   _____

5. **Checking Your Hypothesis** An electromagnet with 1000 coils can lift a 100-kg piece of scrap iron. If the scrap yard owner wants to upgrade the magnet to lift 350 kg, how many coils would the electromagnet need?

   _____

   _____

24-3 Physics Lab

Apply

1. Some security systems use an electromagnet to prevent a door from opening when a building is unoccupied. The electromagnet attracts the door itself if it is made from a magnetic material or a plate of a magnetic material attached to the door. These electromagnets exert thousands of newtons while using a current of several hundred milliamperes. Describe the construction of these electromagnets.

   _____

   _____

   _____

   _____

   _____

   _____

   _____

   _____

25-1

Physics Lab

Electromagnetic Induction 1

In 1831, Michael Faraday discovered that, when a conductor moves in a magnetic field in a direction that is not parallel to the field, an electric current is induced in the conductor. The strongest current is generated when the conductor moves perpendicular to the magnetic field. This process of generating an electric current is electromagnetic induction, and the current produced is induced current. Current is produced only when there is relative motion between the conductor and the magnetic field; it does not matter which moves.

Procedure

1. Attach the single loop of wire to the galvanometer, as shown in Figure A. Thrust one of the bar magnets through the coil. Record your observations in Table 1.

2. Move the galvanometer connections to the 25-turn coil of wire. Thrust the magnet into the coil. Record your observations in the table.

3. Move the galvanometer connections to the 100-turn coil of wire. Thrust the magnet into the coil. Record your observations in the table.

Figure A

Thrust the bar magnet through a single loop of wire.

4. With the galvanometer connected to the 100-turn coil of wire, thrust the north pole of the magnet into the coil. Observe the direction of movement of the galvanometer needle. Pull the magnet out of the coil. Observe the direction of movement of the galvanometer needle. Thrust the south pole of the magnet into the coil. Pull it back out. Note the movement of the galvanometer needle. Record your observations in the table.

5. Tape or hold two bar magnets together in parallel, so that like poles are aligned. Thrust them into the 100-turn coil of wire. Observe the deflection of the galvanometer needle. Slowly pull the two magnets out of the coil. Try different rates of speed for pushing the magnets into and out of the coil and compare the deflections of the galvanometer needle. Record your observations in the table.

Objectives

- **Demonstrate** the generation of electric current from a moving magnetic field.

- **Observe** the relationship between magnetic field strength and current strength.

- **Infer** the relationship between magnetic field polarity and electric current direction.

- **Compare** electric fields generated by static charges to those generated by magnetic induction.

Materials

- galvanometer with zero in center of scale
- 1-turn coil of wire
- 25-turn coil of wire
- 100-turn coil of wire
- 2 bar magnets
- connecting wires

25-1 Physics Lab

6. Place the magnets in the 100-turn coil. While the magnets are stationary, does the galvanometer needle deflect? Move the coil back and forth while the magnets are stationary. Observe the motion of the galvanometer needle. Record your observations in the table.

Data and Observations

Table 1	
	Observations of Needle Deflection
Magnet into 1-turn coil	
Magnet into 25-turn coil	
Magnet into 100-turn coil	
Magnet into and out of coil	
Two magnets into 100-turn coil, different velocities	
Two magnets into 100-turn coil: stationary and moving coil	

25-1 Physics Lab

· · · · · · · · · · · · · · · ·

Analysis and Conclusions

1. In step 4, why did the galvanometer needle deflect in one direction when the magnet went into the coil and in the opposite direction when the magnet was pulled back out?

_____

_____

_____

_____

2. Summarize the factors that affect the amount of current and *EMF* induced by a magnetic field.

_____

_____

_____

_____

3. An equation for the electromotive force induced in a wire by a magnetic field is $EMF = BLv$, where B is magnetic induction, L is the length of the wire in the magnetic field, and v is the velocity of the wire with respect to the field. Explain how the results of this experiment substantiate this equation.

_____

_____

_____

_____

_____

4. What happens when a conducting wire is held stationary in or is moved parallel to a magnetic field? Give a reason for your answer.

_____

_____

_____

5. Compare the induced electric field generated in this experiment to the electric field caused by static charges.

_____

_____

_____

25-1 Physics Lab

Extension and Application

1. The equation $EMF = BLv$ applies only when the wire moves perpendicular to the magnetic field lines. A more general form of the equation would have to account for movement at an angle to the magnetic field lines. Find the induced EMF of a wire 0.40 m long that moves through a magnetic field of magnetic strength 0.75×10^{-2} T at a speed of 5.0 m/s. The angle θ between v and B is 45°. Show the equation you will use to solve for EMF and show all your calculations.

2. Write a statement that describes the effect on EMF of the angle θ between v and B.

   _____

   _____

3. As a conducting loop rotates in an external magnetic field, the induced current alternately increases and decreases. Describe the output produced and predict the shape of a graph that plots current against time.

   _____

   _____

   _____

4. Figure B shows a transformer. This device changes the AC voltage of the primary coil by inducing an increased or decreased EMF in the secondary coil. The value of the secondary voltage depends on the ratio of the number of turns of wire in the two coils. How can a magnetic field move across the secondary coil and induce the EMF? Why does this device operate only on alternating, not direct, current?

Figure B

A transformer has two coils of insulated wire wound around an iron core.

   _____

   _____

   _____

   _____

25-2

Physics Lab

Electromagnetic Induction 2

Electromagnetic induction generates an electric current from the relative motion of a conductor in a magnetic field. The third right-hand rule indicates the direction of the induced current. Hold your right hand flat with your thumb pointing in the direction in which the wire is moving and your fingers pointing in the direction of the magnetic field. The palm of your hand points in the direction of the conventional (positive) current.

Michael Faraday invented the electric generator, which converts mechanical energy into electric energy. A generator has a large number of wire loops wound about an iron core, called the armature, which is in a strong magnetic field. As the armature turns, the wire cuts through the magnetic field and an *EMF* is induced. The *EMF* = *BLv*, where *B* is magnetic induction, *L* is the length of wire rotating in the field, and *v* is the velocity with which the loops move through the magnetic field. An electric motor is the opposite of a generator. In an electric motor, the magnetic field of an electric current, applied to the coil, turns the armature in a magnetic field. Alternately attracted and repelled by fixed magnets, the armature spins, using a magnetic force to convert electric energy to mechanical energy.

Procedure

1. Wind two 120-turn coils of enameled magnet wire around a 100-mL beaker, a toilet-paper tube, or a paper-towel tube. Leave a 15-cm-long wire lead at each end of the coil, coming out about 4 cm apart on the same side of the coil, as shown in Figure A. After winding each coil, carefully slide it off the coil form and wrap several small pieces of tape around the coils so that they maintain their shape.

Objectives

- **Demonstrate** the mechanical behavior of a coil of wire in a magnetic field.
- **Observe** the behavior of two connected coils of wire in magnetic fields.
- **Explain** the differences between a generator and a motor.
- **Infer** Lenz's law from the observations.

Materials

- 100-mL beaker
- enameled magnet wire, 60 m
- galvanometer with zero in center of scale
- ring stand
- 2 ring-stand clamps
- masking tape
- 2 horseshoe magnets
- fine sandpaper
- connecting wires

Figure A

The leads should be about 15 cm long and about 4 cm apart on the coil. Use small pieces of tape to hold the coil together.

25-2 Physics Lab

2. Carefully sand 1 cm of each end of the enameled wires to expose clean copper.

3. Set up a ring stand with a clamp approximately 20 cm above the bottom of the stand. Place the second clamp next to the first one, but pointing in the opposite direction. Hang the coils from the clamps, as shown in Figure B, using a piece of tape to secure the wire leads to the clamps. Position the horseshoe magnets as shown in the figure. Adjust the height of the clamps so that one bar of each horseshoe magnet is in the center of each coil and projects several centimeters through the coil.

Coil leads

Clamp

Coil

Horseshoe magnet

Ring stand

Figure B

4. Connect one set of coil leads with wire to the galvanometer. Swing this coil on the horseshoe magnet and observe the galvanometer needle. Swing the coil faster. Try adding a stronger magnet or put two magnets side by side with like poles together. Swing the coil again. Record your observations in item 1 of Data and Observations.

5. Use the right-hand rule to determine the direction of the current to the galvanometer when the first coil is pushed toward the magnet. Label the positive lead. Attach the galvanometer to the leads of the other coil. Swing this coil. Again use the right-hand rule to determine the current to the galvanometer when you push the second coil toward the magnet. Label the positive lead.

6. Disconnect the galvanometer. Swing one coil and observe its motion. With a wire, connect the leads that you marked positive. Connect the other two leads with a piece of wire.

7. Start one coil swinging. Observe the entire system. Record your observations in item 2 of Data and Observations. How does the speed of the swinging coil compare to its motion when it is not attached to any other components? Record your observations in item 3.

8. Unhook one set of connecting wires and insert the galvanometer in series with the two coils. Start one coil swinging. Observe the galvanometer. Record your observations in item 4.

Data and Observations

1. Observations of swinging coil on magnet:

_____

_____

_____

_____

2. Observations of swinging coil attached to other coil:

_____

_____

_____

_____

25-2 Physics Lab

3. Observations of unhooked swinging coil compared to its motion in the circuit:

_____

_____

_____

4. Observations of system with galvanometer:

_____

_____

_____

_____

_____

_____

Analysis and Conclusions

1. Use your observations from item 1 to summarize the factors that affect the strength of the induced current.

_____

_____

_____

2. Use your observations from item 2 to explain the motion of the coils.

_____

_____

_____

_____

_____

3. Compare the rates of swinging of the coil that you observed in steps 6 and 7. How can the difference be explained?

_____

_____

_____

_____

_____

25-2 Physics Lab

4. Which coil acts like a generator, and which coil acts like a motor? Give a reason for your answer.

_____

_____

_____

_____

_____

Extension and Application

1. When a load is applied to or increased on a generator, why is it more difficult to keep the generator turning? You may have felt this effect when you turned on bicycle lights powered by a pedal-driven generator.

_____

_____

_____

_____

_____

_____

_____

26-1 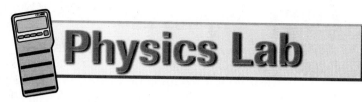 Physics Lab

Mass of an Electron

Objectives

- **Observe** the deflection of electrons moving through a magnetic field.
- **Determine** the mass-to-charge ratio for an electron.
- **Determine** the mass of an electron.
- **Calculate** the experimental error relative to the accepted value.

J. J. Thomson was the first to measure the ratio of mass to charge for an electron. He observed the deflection of a beam of electrons passing through combined electric and magnetic fields. In his experiment, the electric and magnetic fields exerted forces perpendicular to the direction of the electrons' motion. He applied a fixed electric field and adjusted the magnetic field until the electrons traveled a straight path (zero deflection). By equating the forces due to the fields, Thomson could calculate the ratio of mass to charge.

In this experiment, you will follow a process like Thomson's to balance the forces on electrons and to determine the ratio of mass to charge. You will be using a mass-of-the-electron apparatus—a 6E5-style tube. The cathode is in the center of the tube under a circular metal cap. Electrons emitted from the cathode accelerate horizontally toward the large conical-shaped anode that nearly fills the top of the tube. The anode is coated with a fluorescent material that glows when electrons strike it.

In the absence of any external electric or magnetic fields, the electrons falling on the anode produce the pattern shown in Figure A1. If you put the tube in an air-core solenoid that has current flowing in it, the electrons are subjected to a constant magnetic field acting perpendicular to their direction of motion. Since the electrons are moving at a fairly uniform speed, the situation is that of a moving charged particle subjected to a constant force. As a result, the electrons move in an arc. The pattern observed on the anode of the tube will be like the one in Figure A2.

To find the radius of curvature of the electron path, you will compare the tube pattern with a round object such as a wooden dowel. You may have to adjust the coil current or the anode voltage to get a good match.

Figure A

Undeflected pattern

Deflected pattern

26-1 Physics Lab

When an electron (mass $= m$) is acted upon by a magnetic force in a direction perpendicular to its motion, it travels in the arc of a circle with radius r. The centripetal force, F_c, produces circular motion with an acceleration v^2/r, giving

$$F_c = \frac{mv^2}{r}.$$

The force exerted by the magnetic field is equal to Bqv. When the two forces are equal,

$$Bqv = \frac{mv^2}{r}.$$

Solving for v yields

$$v = \frac{Bqr}{m}$$

and, by squaring both sides,

$$v^2 = \frac{B^2q^2r^2}{m^2}.$$

The electrons accelerate through a potential difference in the tube. The energy that each electron acquires as it accelerates through a potential difference is qV, where q is the charge carried by an electron in coulombs, and V is the potential difference in volts. Since $1\ V = 1\ J/C$, qV is expressed in joules. The energy acquired by the electron is kinetic energy, $\frac{mv^2}{2}$. Thus,

$$qV = \frac{mv^2}{2},$$

where v is the velocity the electron acquires as it moves through a potential difference, V. If the particle then passes perpendicularly through a magnetic field, B, the velocity is the same v found from equating the electric and magnetic forces. Therefore, if we substitute for v^2,

$$qV = \frac{mv^2}{2}\ \text{becomes}$$

$$qV = \frac{mB^2q^2r^2}{2m^2},$$

and, by rearranging terms,

$$\frac{m}{q} = \frac{B^2r^2}{2V}.$$

An electron carrying a charge q accelerated by a known voltage V, entering a magnetic field of magnitude B, will travel in a circular path of radius r, which can be measured to determine the mass-to-charge ratio. Since the charge q, on an electron is known, 1.6×10^{-19} C, you can determine the mass of an electron m, from

$$m = q\left(\frac{m}{q}\right),$$

where $\frac{m}{q}$ is the ratio you calculated earlier.

26-1 Physics Lab

Materials

mass-of-the-electron apparatus

DC power source, 0–5 A

DC power source, 90–250 V

DC or AC filament power supply, 6V

air-core solenoid

connecting wires

ammeter

3 dowels, different diameters

voltmeter

CBL unit

link cable

graphing calculator

magnetic field sensor

CBL-DIN adapter

Procedure

1. Connect the CBL unit to the graphing calculator, using the unit-to-unit link cable. Connect the magnetic field sensor to the CH1 port of the CBL unit. Turn on the CBL unit and the graphing calculator. Select SET UP PROBES from the MAIN MENU. Enter 1 as the number of probes. Select MORE PROBES from the SELECT PROBE menu. Select the MAGNETIC FIELD probe from the choices. Enter 1 as the channel number for the probe. For CALIBRATION, select USE STORED. From the MG FIELD SETTING menu, select HIGH(MTESLA). From the MAIN MENU, select OPTIONS. Then choose ZERO SENSOR from the PHYSICS OPTIONS menu. Select channel 1 from the SELECT CHANNEL choices. Hold the magnetic field sensor vertically with the dot facing either north or south. Observe the CBL display. When the display shows a constant value, press TRIGGER on the CBL unit to zero the sensor. Select COLLECT DATA from the MAIN MENU. Select the MONITOR INPUT option from the DATA COLLECTION menu. The graphing calculator will begin displaying magnetic field values.

2. Wire the mass-of-the-electron apparatus as shown in Figure B. The tube should be premounted and supplied with color-coded wires that match the figure. Have your teacher check your wiring.

Figure B

3. Turn on the low-voltage filament power supply. Wait about 30 s before applying the high-voltage plate supply. Set the plate voltage between 130 and 250 V. Inspect the resulting pattern. It should match the pattern shown in Figure A1.

Figure C

26-1 Physics Lab

4. Connect the air-core solenoid to the power source as shown in Figure C. **CAUTION:** *The tube is fragile and can be easily broken.* Carefully place the coil over the tube. Turn on the power supply to the coil. Increase the current to 3–4 A. The tube pattern should now look like the pattern in Figure A2. Turn off the coil current.

5. Carefully insert one of the dowels, or another round, nonmetal object such as a pencil, into the coil and rest the end of it on top of the tube. Apply power to the coil and adjust the current until the curvature of the tube pattern is close to the curvature of the dowel. (The pattern will not match exactly.)

6. Record in Table 1 the radius of the dowel, the accelerating voltage, and the coil current that causes the deflection.

7. Repeat steps 4–6, making several more trials with dowels of other sizes. If you want, change the accelerating voltage.

8. Place the magnetic field probe against the center of the solenoid with the flat side with the dot facing out from the coil. Adjust the coil current to the values recorded in Table 1, and record the corresponding coil magnetic induction value in Table 1.

Data and Observations

Table 1				
Trial	Coil Current, I (A)	Magnetic induction, B (T, N/A·m)	Potential Difference, V (V)	Radius, r (m)
1				
2				
3				

Table 2				
Trial	B^2 (N/A·m)	r^2 (m²)	Charge on one electron, q (C)	mass/charge (kg/C)	Mass of electron (kg)
1					
2					
3					

Physics: Principles and Problems

26-1 Physics Lab

Analyze and Conclude

1. In Table 2, record the known value of q. Calculate B^2, r^2, $\frac{m}{q}$, and the mass of an electron. Enter the results in Table 2.

2. Determine the average mass for one electron.

3. Compare the average value of the mass of an electron with the accepted value and find the relative error.

4. Despite the error in your measured values, what is the significance of these experimental results?

_____

_____

Extension and Application

1. The magnetic field at the surface of Earth is about 5×10^{-5} T. Instead of the air-core solenoid, could Earth's magnetic field be used to deflect the beam of electrons? If so, propose a design for an experiment to accomplish this and describe the required components and procedure.

_____

_____

_____

_____

_____

_____

27-1

Physics Lab

Planck's Constant

While studying radiation emitted from glowing, hot material, Max Planck assumed that the energy of vibration, E, of the atoms in a solid could have only specific frequencies, f. He proposed that vibrating atoms emit radiation only when their vibrational energy changes, and that energy is quantized, changing only in multiples of hf. This relationship is $E = nhf$, where n is an integer and h is a constant. A light-emitting diode, or LED, is a modern application of this phenomenon.

An LED is made of a semiconductor wafer that has been *doped* with two types of impurities. When doped with impurities that have loosely bound electrons, a semiconductor donates those electrons and is designated an *n*-type material. A semiconductor doped with acceptor impurities has *holes* that collect electrons and is designated a *p*-type material. When *n*-type and *p*-type impurities are combined in a semiconductor, a *pn* junction is formed that acts like a vacuum-tube diode, permitting current to flow in only one direction. If a direct current is applied in the circuit when the *pn* junction is reverse biased, the diode has a very high resistance and does not conduct current. When the *pn* junction is forward biased, there is a very low resistance, and a large current can flow through the diode. An LED emits light when current moves across a forward-biased *pn* junction. LEDs can produce light having a wide variety of wavelengths, ranging from the far-infrared region to the near-ultraviolet. The different wavelengths of visible light in LEDs are produced by varying the type and amount of impurity added to the semiconductor crystal.

Objectives

- **Demonstrate** the relationship between the voltage applied to an LED and the wavelength of the light given off by that LED.
- **Model** and **interpret** the relationship between voltage and wavelength for each color of LED.
- **Calculate** the value of Planck's constant for each color of LED.
- **Determine** the experimental error.

When current moves across a forward-biased *pn* junction, free electrons from the *n*-type material are injected into the *p*-type material, as shown in Figure A. When these carriers recombine, energy is released. The energy released from vibrations in the crystal lattice can be in the form of light or heat. The proportions of heat and light produced are determined by the recombination process taking place. As Planck proposed, the energy produced is $E = hf = hc/\lambda$, where E is the energy in joules, h is Planck's constant, c is the speed of light (3.0×10^8 m/s), f is the frequency of the emitted light, and λ is the wavelength of the emitted light. In an LED, energy is supplied from a battery or DC power supply. Electric energy

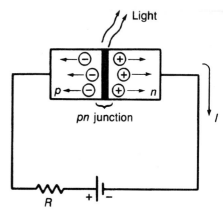

Figure A

A forward-biased, LED *pn* junction.

27-1 Physics Lab

Forward current

Knee

Voltage

Figure B

Materials

green, red, and yellow
   LEDs

2 1.5-V batteries or 3-V
   power supply

battery holder

22-Ω resistor

1000-Ω potentiometer

connecting wires

ammeter, 0–50 mA DC

voltmeter, 0–5 VDC

knife switch

supplied to the charge is $E = qV$, where E is the energy in joules, q is an elementary charge $(1.6 \times 10^{-19}$ C), and V is the energy per charge in volts. These two relationships for E are equal.

$$qV = \frac{hc}{\lambda}$$

Rearranging terms and solving for h yields the following.

$$h = \frac{qV\lambda}{c}$$

Figure B shows a typical curve for a graph of current versus voltage in a forward-biased diode. The point at which recombination begins producing a significant amount of light, compared to heat, is the knee. At the knee, the resistance drops sharply, and the current increases rapidly within in the diode.

Procedure

1. Wire the circuit shown in Figure C. **CAUTION:** *Handle the LEDs carefully; their wire leads are fragile and will not tolerate much bending.* Be sure to observe the correct polarity for the meters. Have your teacher inspect your circuit before you continue.

2. Record in Table 1 the colors and the wavelengths of the LEDs supplied by your teacher.

3. **CAUTION:** *At no time during the experiment should the current to the LED exceed 25 mA.* Rotate the potentiometer control to its center position. Close the switch and observe the voltage on the voltmeter. Slowly adjust the voltage to approximately 2.0 V. If the LED is not glowing, open the switch and reverse the LED leads. The potentiometer forms a voltage divider across the power supply to supply voltages from 0 to 3 V. Begin at 1.50 V, and increase in increments of 0.05–0.10 V until the current is less than or equal to 25 mA. Collect current readings for the various voltage settings and record your data in Table 2. When you have taken the last reading near or equal to 25 mA, turn off the current to the circuit.

4. Replace the LED with one of another color and repeat step 3. Collect data for all of the LEDs.

Figure C

Schematic diagram of a
circuit to measure voltage
and forward current across
an LED

27-1 Physics Lab

Data and Observations

Table 1		
LED	LED color	Wavelength (nm)
1		
2		
3		

Table 2					
LED 1		LED 2		LED 3	
Voltage (V)	Current (mA)	Voltage (V)	Current (mA)	Voltage (V)	Current (mA)

27-1 Physics Lab

Analysis and Conclusions

1. Make a graph that plots voltage on the horizontal axis and current on the vertical axis. Plot each set of LED data separately on one set of axes, and label each curve. As the curve moves from zero current through the knee to a linear relationship, determine the point where the graph becomes linear. Hint: The current will be 5–10 mA. For each LED curve, find the corresponding voltage for this point. At this voltage, recombination produces a significant amount of light. List below the voltages for each LED.

   Red voltage: _____

   Yellow voltage: _____

   Green voltage: _____

2. Calculate the value of Planck's constant for each LED. Show your work.

3. Compute the relative error for Planck's constant for each trial, using $h = 6.626 \times 10^{-34}$ J·s as the accepted value.

4. What approximate value for V might be expected for an LED that produces blue light and for one that produces infrared light?

   _____

   _____

Extension and Application

1. Considering what you have learned in this experiment, would it be possible to construct an LED that generates white light? Give a reason for your answer.

   _____

   _____

   _____

2. What advantages and/or disadvantages are there to using LEDs rather than a regular incandescent lightbulb?

   _____

   _____

27-2

Physics Lab

The Photoelectric Effect

Objectives

- **Measure** the kinetic energy of photoelectrons for various colors of light.

- **Model** and **interpret** the relationship between light frequency and kinetic energy.

- **Calculate** the threshold frequency for a photocell.

Photoelectrons are ejected from the cathode of a phototube only if the frequency of the incident electromagnetic radiation is above a certain minimum value, called the threshold frequency. The phototube is a round glass tube containing a semicircular cathode, a small, centrally located anode wire, and no air. Light shining on the cathode causes emission of photoelectrons, which travel through the tube and strike the anode. As shown in Figure A, an external wire connects the anode to an amplifier circuit. Current through this wire, detected by the microammeter, indicates when a photoelectron current is present. A potentiometer located on the photoelectric-effect module allows adjustment of the applied voltage. The voltage is adjusted until it balances the kinetic energy of the photoelectrons. This balance is indicated by the absence of a photoelectron current, and the voltage required to stop the current is the stopping voltage. The energy associated with a potential difference is $E = qV$, where E is the energy in joules, q is an elementary charge with $q = 1.6 \times 10^{-19}$ C, and V is the voltage in volts.

Phototube

To current amplifier/microammeter

Voltmeter (stopping voltage)

Potentiometer to adjust stopping voltage

Figure A

Albert Einstein explained the photoelectric effect in terms of the photoelectric equation, $K = hf - hf_o$, where h is Planck's constant, f is the frequency of incident light, f_o is the threshold frequency, and K is the kinetic energy of the emitted photoelectrons.

Procedure

1. Set up the photoelectric-effect module, voltmeter, microammeter, and current amplifier, according to the instructions supplied with the equipment. Make calibration adjustments specified in the manufacturer's instructions.

2. Place the light source in front of the window opening to the photoelectric tube. Put a colored filter over the window. Check that the filter completely covers the opening and prevents stray light from entering the window.

3. Slowly adjust the stopping voltage until the photocurrent drops to zero. Record in Table 1 the filter color and its stopping voltage.

Materials

- photoelectric-effect module with amplifier

- voltmeter, 0–5 VDC

- microammeter, 0–100 μA

- colored filters: red, green, blue, yellow, and violet

- 40-W incandescent lightbulb in socket

27-2 Physics Lab

4. Replace the filter with one of another color. Repeat step 3 with this filter and then with all the other filters.

5. Determine the energy associated with each stopping voltage by multiplying the stopping voltage by 1.6×10^{-19} C. Record the values in Table 1.

Data and Observations

Table 1				
Filter color	Stopping voltage (V)	Energy (J)	Wavelength (m)	Frequency (Hz)

Analysis and Conclusions

1. Record in Table 1 the wavelength for each color. Use the wavelengths provided by your teacher, or find the wavelengths in a reference book. Calculate the frequency associated with each wavelength, using $c = f\lambda$, where $c = 3.0 \times 10^8$ m/s.

2. The energy equivalent to the stopping voltage is the energy needed to match the kinetic energy of the photoelectrons. Plot a graph with kinetic energy of photoelectrons, in joules, on the y-axis and the frequency of incident light on the x-axis. Draw the straight line that best fits your data.

3. What is the value and significance of the line that intersects the x-axis at a point other than 0, 0?

_____

_____

_____

_____

4. Determine the slope of the line. What is its significance?

_____

_____

_____

_____

Extension and Application

1. Photographic darkrooms use red safelights while black-and-white prints and some special black-and-white films, such as X-ray film, are processed. The safelights can be used because they do not further expose the film or paper. Why are safelights red?

28-1

Physics Lab

Spectra

Objectives

- **Observe** the emission spectra of various elements in gas state.

- **Compare** the intensity of spectral lines for different elements.

- **Analyze** a bright-line spectrum.

- **Recognize** an element from its emission spectrum.

When solids glow with heat, their atoms produce a continuous spectrum. Substances that are vaporized by heating emit light characteristic of the elements in the substance. For example, a solution of sodium chloride placed on a platinum wire and held in a flame emits a bright yellow light. Another way to produce a spectrum is to apply high voltage across a gas-filled glass tube. Gas atoms that are under low pressure and excited by an electrical discharge give off light in characteristic wavelengths. When the emitted light passes through a spectroscope, it breaks into its constituent components for analysis. A gas viewed through a spectroscope, such as the one shown in Figure A, forms a series of bright lines known as a bright-line or emission spectrum.

Since each element produces a unique bright-line spectrum or pattern, spectroscopy is valuable for detecting the presence of elements. Sodium, for example, gives off bright yellow light that appears in the spectroscope as two adjacent bright lines of yellow. A gas is identified by comparing the wavelengths of its emission spectrum to the spectrum produced by a known gas.

Figure A

Spectrum tube

Spectroscope

28-1 Physics Lab

Materials

spectroscope

spectrum-tube power
 supply

spectrum tubes: argon,
 bromine, carbon
 dioxide, chlorine,
 helium, hydrogen,
 krypton, mercury
 vapor, neon, nitrogen,
 oxygen, and xenon

40-W incandescent
 lightbulb and socket

thermal mitt

Procedure

CAUTION: *The power supply and spectrum tubes use several thousand volts. Do not touch the spectrum-tube power supply or spectrum tubes when power is applied. Use thermal mitts to handle the tubes.*

1. Look through a spectroscope at an incandescent lightbulb. The spectrum should appear when the slit in the spectroscope is pointed just off center of the glowing filament. Practice moving the spectroscope until you see a bright, clear image.

2. Verify that the power to the spectrum-tube power supply is turned off. Insert one of the spectrum tubes into the sockets. Helium or hydrogen is a good first choice.

3. Turn on the power supply. Darken the room but leave enough background lighting to illuminate the spectroscope scales. If an exposed window is the background lighting source, point the spectroscope away from the window, since daylight will affect the observed gas spectrum. Adjust the spectroscope until the brightest image is oriented on your scale. Some of the spectrum tubes produce light so dim that you must be very close to them to get good observations of the spectral lines. Record in Table 1 the bright lines of the observed spectrum. The width and darkness of your lines should reflect your observations. Turn off the power supply and, using a thermal mitt, carefully remove the hot spectrum tube. Your teacher will tell you where to put these hot tubes so that other students don't accidentally touch them and get burned.

4. Repeat steps 2 and 3, using all the other spectrum tubes. Record your observations in Table 1.

Data and Observations

28-1 Physics Lab

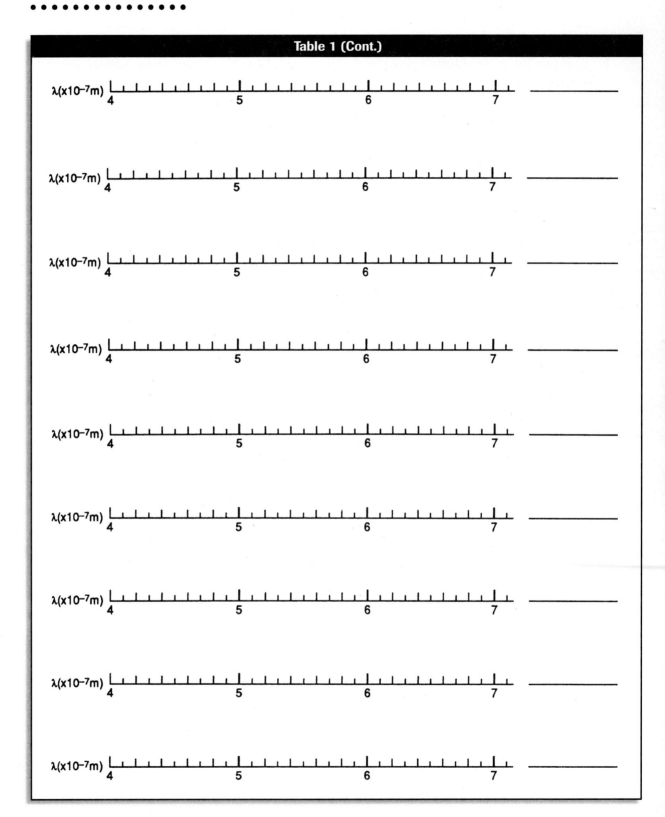

Table 1 (Cont.)

λ(x10⁻⁷m)   4   5   6   7   _____

λ(x10⁻⁷m)   4   5   6   7   _____

λ(x10⁻⁷m)   4   5   6   7   _____

λ(x10⁻⁷m)   4   5   6   7   _____

λ(x10⁻⁷m)   4   5   6   7   _____

λ(x10⁻⁷m)   4   5   6   7   _____

λ(x10⁻⁷m)   4   5   6   7   _____

λ(x10⁻⁷m)   4   5   6   7   _____

λ(x10⁻⁷m)   4   5   6   7   _____

28-1 Physics Lab

Analysis and Conclusions

1. Compare the color of light emitted by the spectrum tube to that observed through the spectroscope. Explain any differences.

_____
_____
_____

_____

2. Compare the intensities of the observed spectral lines for each element.

_____

_____

3. Observe a fluorescent lightbulb with the spectroscope. While a continuous spectrum will be visible, you will also see a bright-line spectrum. Compare this spectrum with those of the gases observed in this activity and identify the gas in a fluorescent lightbulb.

_____

_____

Extension and Application

1. A pulse of high-voltage electricity passed through a tube filled with xenon gas creates the electronic flash that is used for photography. Examine the spectrum of xenon and compare it with other gases. Why is xenon used rather than some other gas?

_____
_____
_____

2. Describe the uses of a spectroscope in the science of astronomy.

_____
_____
_____

3. Suggest a practical use for a spectroscope in the laboratory.

_____
_____
_____

Physics: Principles and Problems

28-2

Physics Lab

Measuring Electron Energy Changes

Objectives

- **Observe** the characteristic emission spectrum of sodium.
- **Infer** the energy absorption and emission of atoms.
- **Solve** trigonometric equations.
- **Determine** the wavelength of light emitted by sodium.

Materials

apron

laboratory burner

lighter

2 metersticks

tape

modeling clay

goggles

diffraction grating

cardboard with slit

white paper

sample of anhydrous sodium carbonate salts

When you heat a substance until it vaporizes, the elements in it emit light at characteristic wavelengths. Electrons absorb energy from the flame and move to higher orbitals. When these electrons return to a lower orbital, they give off energy in the form of light. Because energy changes occur in fixed steps and each element has a unique electron structure, each element has a unique emission spectrum. By measuring the wavelengths of light emitted during a flame test, you can identify an element.

You will use a diffraction grating in this experiment. A diffraction grating has parallel fine grooves, sometimes as many as 20 000 lines/cm. When you look through the grating at light passing through a slit, the image of the slit appears at a location that is proportional to the wavelength of the light and the number of grooves on the grating. You used this method for the hand-held spectroscope in Lab 28-1, except that the spectroscope is calibrated in wavelengths.

Procedure

1. Fasten a meterstick to the table using modeling clay as in Figure A. Use clay to attach a piece of cardboard with a slit, so that the slit lines up with the center of the meterstick.

2. From the center of the meterstick, position another meterstick perpendicularly and attach it to the table with clay. Place a diffraction grating, mounted on a card or slide mount, 1.00 m from the slit. (This is distance S.)

3. Place a laboratory burner behind the cardboard. Be sure the cardboard will not be too close to the flame. Put on your goggles and apron now, and tie back long hair.

Figure A

28-2 Physics Lab

4. Tape a sheet of paper on the wall behind the burner on one side of the slit. Light the burner and adjust the flame until it is totally blue. Check that you can observe the flame through the slit.

5. Obtain a covered bottle of sodium carbonate.

6. You and your partners will need to work together to collect the data.

   • Student A looks through the diffraction grating.

   • Student B marks the paper where student A sees a spectral line.

   • Student C supplies the flame with powdered salt.

7. Student A observes the burner flame through the diffraction grating and slit. Student C shakes a covered bottle of salt and then opens it close to the collar (air inlet) of the burner. Student B marks the location of a spectral line on the paper on the wall. Note in the table the color of the spectral line that Student A observes.

8. Measure the distance from the slit to the sighted spectral line. Record this distance (D) in the table.

Data and Observations

Table 1			
Spectral Line			
Color	Distance D	Distance S	Lines/cm n
Sodium			

Analysis and Conclusions

1. Use a calculator and the Pythagorean theorem to determine the distance H from the diffraction grating to the spectral line.

$$H = \sqrt{D^2 + S^2}$$

= _____

2. The sine of angle AOB of triangle AOB (see Figure A) is

$$\sin \theta = \frac{D}{H}.$$

Use your calculator to find $\sin \theta$.

$\sin \theta =$ _____

28-2 Physics Lab

3. Find out the number of lines per centimeter in the diffraction grating n from your teacher. The distance between the lines in the spectrum is $d = 1/n$. Calculate the distance between the lines in the spectrum.

$d =$ _____

4. According to the Bragg equation, the wavelength of light emitted by an element is $d \times \sin \theta$. Calculate the wavelength in nanometers of the light emitted by the energized sodium atoms.

$d \times \sin \theta =$ _____

Extension and Application

1. Suppose you repeated the experiment with an element that emits blue light. Will the distance D in the experiment be greater or less than what you observed for sodium? Give a reason for your answer.

_____

_____

_____

_____

_____

_____

2. Many of the newer lights used to illuminate roads and highways emit light that appears either yellow or blue. The color comes from an element that the heat of the lamp vaporizes. Yellow lamps use sodium, and blue lamps use mercury. These lamps use the same amount of electricity to produce more light than ordinary lamps. Explain why these lamps are more efficient.

_____

_____

_____

_____

_____

_____

29-1

Physics Lab

Semiconductor Properties

The diode is the simplest type of semiconductor device. It is made of a semi-conductor material, such as germanium or silicon, that is specially doped with two types of impurity. The *n*-type end of the diode is doped with a donor impurity, such as arsenic, that has loosely bound electrons. The *p*-type end is doped with an acceptor impurity, such as gallium, that has "holes" for electrons. Figure A is a schematic representation of the diode.

Diode   **Figure A**   Diode symbol

When a small direct current is applied to a diode in one direction, called reverse bias, the diode has a very high resistance and conducts almost no current. When current is applied in the other direction, called forward bias, the diode has a low resistance, and a large current flows through the diode. All semiconductor diodes have a voltage rating that determines how much voltage can be applied to the diode in the reverse-bias direction before the diode starts to conduct a large current in that direction. When this voltage rating, usually hundreds of volts, is exceed-ed, the diode no longer works properly and usually shorts out. In this experiment, you will forward- and reverse-bias a diode and measure the current as a function of the applied voltage. The voltage at which a diode begins to conduct current in the forward-bias direction depends on the material used to make the diode. Silicon diodes begin to conduct current at about 0.6 V, and germanium diodes at about 0.2 V.

Objectives

- **Demonstrate** that diodes conduct current in one direction.
- **Measure** the current flowing through a diode for various voltages across the diode.
- **Analyze** the relationship between voltage and current for a diode.
- **Observe** the properties of a diode when connected to a source of alternating current.

Materials

- 1N4004 diode (rectifier) or equivalent
- battery, 1.5–3-V
- 10-Ω resistor
- 1000-Ω resistor
- 1000-Ω potentiometer
- connecting wires
- switch
- AC power supply, 0–5 VAC
- CBL unit
- graphing calculator
- dual-channel amplifier
- current and voltage probe
- 2 CBL-DIN adapters
- link cable

Procedure

1. Construct the circuit shown in Figure B. The voltage is applied to the *pn* junction so that the *p*-type end is positive with respect to the *n*-type end. Note that a band or ring around one end of the diode indicates the *n*-type end.

2. Connect the CBL unit and graphing calculator with the unit-to-unit link cable. Connect a CBL-DIN adapter to CH1 and CH2 ports on the CBL unit. On the amplifier box, connect the DIN1 lead to the CBL port CH1 and the DIN2 lead to port CH2. Connect the cur-rent1 probe to PROBE1 port on the amplifier box, and connect the voltage1 probe to PROBE2 port of the amplifier box. Turn on the CBL unit and your graphing calculator. Start the PHYSICS program.

3. From the MAIN MENU, select SET UP PROBES. Enter 2 as the num-ber of probes. Select MORE PROBES from the SELECT PROBE

Copyright © by Glencoe/McGraw-Hill

29-1 Physics Lab

menu. Select the C-V CURRENT probe from the list. Enter 1 as the channel number. Select
USE STORED from the calibration menu. Again from the SELECT PROBE menu, choose
MORE PROBES and then select C-V VOLTAGE. Enter channel number 2 for the chan-
nel number. Select USE STORED from the CALIBRATION menu.

4. Select the COLLECT DATA option from the MAIN MENU.
From the DATA COLLECTION menu, select MONITOR INPUT.
Close the switch. Turn the potentiometer and observe the voltage displayed on the calculator. Slowly increase the
voltage from 0 V to 0.8 V, recording the corresponding
current readings in Table 1 in the Forward-bias diode.
Press the "+" to quit monitoring
data on the graphing calculator.
Open the switch to turn off power
to the circuit.

5. Reverse the diode so that
the positive end of the
diode, the *p*-type end, is
attached to the current probe.
Repeat step 4, collecting data for
the Reverse-bias diode.

Figure B

6. Dismantle your circuit and assemble the circuit shown in Figure C. Connect an AC source from a
power supply or a small AC transformer adapter, such as used to power a small tape recorder or
portable telephone, as shown in Figure C. **CAUTION:** *Under no circumstances should the diode be
connected to a standard electrical outlet.* Plug the voltage probe into the CBL unit in CH1. From the
MAIN MENU, select SET UP PROBES. Enter 1 as the number of probes. Select MORE PROBES
from the SELECT PROBE menu. Then select C-V VOLTAGE from the SELECT PROBE menu. Enter 1
as the channel number. Use STORED from the CALIBRATE MENU. Select the COLLECT DATA
option from the MAIN MENU. From the DATA COLLECTION menu, select TIME GRAPH. Enter
0.0005 s as the TIME BETWEEN SAMPLES. Enter 99 as the number of samples. Press ENTER. From
the next screen, select USE TIME SETUP.

7. Attach the voltage probe to the AC power source first. Press ENTER to begin data collection. Sketch
the displayed wave shape in Table 2 in the AC wave shape box.

8. Move the voltage probe to the 1000 Ω resistor. Select the COLLECT DATA option from the MAIN
MENU. From the DATA COLLECTION menu, select TIME GRAPH. Enter 0.0005 s as the TIME
BETWEEN SAMPLES. Enter 99 as the number of samples. Press ENTER. From the next screen,
select USE TIME SETUP. Press ENTER to begin data collection. Sketch the displayed wave shape in
Table 2 in the AC wave shape through a diode box.

Figure C

29-1 Physics Lab

Data and Observations

Table 1									
Voltage (V)	0.00	0.10	0.20	0.30	0.40	0.50	0.60	0.70	0.80
Forward-biased diode, current (mA)									
Reverse-biased diode, current (mA)									

Table 2	
Diode with AC power	
AC wave shape	
AC wave shape through a diode	

Analysis and Conclusions

1. On a sheet of graph paper, plot a graph of current versus voltage, with current on the vertical axis. Set up the horizontal axis with both negative and positive voltages and center the vertical axis. The negative voltage values represent the reverse-biased diode, while the positive voltage values represent the forward-biased diode.

2. Compare the graph of current versus voltage with a diode in the circuit to a similar graph with a resistor in the circuit.

   _____

   _____

   _____

   _____

3. What is unique about the current flow through a diode?

   _____

   _____

4. Did you use a germanium or a silicon diode? What data support your answer?

_____

_____

5. Compare the two current wave shapes you observed. Explain any differences.

_____

_____

_____

_____

6. Describe the wave shape of the AC input across the 1000-Ω resistor if the diode is reversed.

_____

_____

Extension and Application

1. The circuit that you constructed with the single diode and an input is called a half-wave rectifier. Conventional current flow is in the direction of the arrow on the diode symbol. A full-wave rectifier circuit, shown in Figure D, is used in electronic equipment, such as computers, televisions, stereo receivers, and amplifiers, to provide DC voltage from the AC power line. Draw the wave shape that you predict would appear across the output load resistor, R_L, for the circuit shown in Figure D.

Figure D

2. Build the circuit shown in Figure D and attach the oscilloscope to the output load resistor. Compare your predicted wave shape to the one you observe.

29-2

Physics Lab

Integrated-Circuit Logic Devices

Objectives

- **Observe** the properties and functions of integrated-circuit logic devices.

- **Obtain** the truth table for several types of logic devices.

- **Compare** different logic functions.

- **Predict** the results of different logic connections.

An integrated circuit, IC, may have dozens to millions of elements, such as transistors, resistors, diodes, and conductors, that together form one or more electronic circuits. The individual components are manufactured on an appropriately doped piece of semiconductor substrate, usually silicon. These components are built up layer by layer on the substrate, and integrated electrically to perform a specific function. After construction, small wires are attached to connecting pads on the IC and run to external metal leads or terminals. The IC is then packaged in a plastic or ceramic container to protect it from moisture and environmental pollutants.

Data entered into a microcomputer or a calculator are manipulated with binary mathematics. In the binary system, all numbers are represented by combinations of zeros and ones. This system can be interpreted by transistor circuitry in ICs. A transistor switch that is on represents a one, and a switch that is off represents a zero. An IC has one or more gates, the basic building block of electronic logic systems. A gate is an electronic switch that controls the current flowing between the terminals. The output of an IC can be monitored using a light-emitting diode (LED) as a logic probe. When a transistor circuit is turned on, the LED probe lights to indicate a logical 1 (a true condition). When the transistor circuit is turned off, the LED does not light, indicating a logical 0 (a false condition).

The binary operations are sometimes called logical operations or Boolean logic: if *a* and *b* then *c*. Adding large numbers or handling more complicated operations, such as multiplication or division, requires many logical operations.

Materials

- breadboard
- 9-V battery
- 9-V battery clip
- connecting wires, 22 gauge
- light-emitting diode (LED)
- 1000-Ω resistor
- CMOS integrated circuits: 4081, 4071, 4001, 4011

Procedure

1. Study your breadboard. A center channel runs lengthwise across the middle of the board. An IC is placed on the board. Pins from one side of the IC fit into holes on one side of the channel, and pins from the other side fit into holes on the opposite side of the channel. Along the outer edges of the breadboard are two rows, or buses, of electrically connected holes that run parallel to the center channel. Insert a lead from each side of the battery to a hole in each row, as shown in Figure A. This procedure creates a row of positive connecting points and a row of negative ones. Between the center channel and each outer bus are columns of electric connecting points. All columns between the center channel and the positive or negative bus are electrically connected. Electric connections to the board are made by carefully inserting exposed wires into the holes of the breadboard.

29-2 Physics Lab

Figure A

Layout of the breadboard

This column is electrically connected.

Red

+

−

9 V

Black

Battery

This row (bus) is electrically connected.

Center channel

This row (bus) is electrically connected.

2. Place the LED in the lower right corner of the breadboard with one lead in one of the holes of the negative bus. Place the other lead in one of the column holes. Connect a 1000-Ω resistor between a hole in the column where the LED is and an adjacent column. Attach a long wire to the column where the 1000-Ω resistor terminates. This wire is your test probe. To check the LED, touch the test probe to a hole along the positive bus. If the LED does not light, unplug it and reverse the holes into which the two wires are inserted. The LED should now light. If it does not, ask your teacher for help.

3. Disconnect the 9-V battery. Handle the ICs carefully because they can be damaged by static electricity. Discharge any excess static electricity by touching a water faucet before handling an IC. Select the 4081 IC, which has four AND gates. An AND gate symbol is shown in Figure B and is represented mathematically by $Z = A \cdot B$, which means Z equals A and B. The schematic representation of the four gates in the 4081 (quad 2-In AND gate) is shown in Figure B.

Figure B

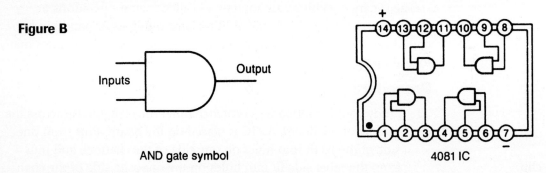

Inputs

Output

AND gate symbol

+

14 13 12 11 10 9 8

1 2 3 4 5 6 7

−

4081 IC

Pin 14 is the positive-voltage input, and pin 7 is the negative-voltage input to power the IC. Look on the top of the IC. The orientation of the IC is usually identified by a semicircular notch on the end near pins 1 and 14 and an index mark, such as a hole or dot, by pin 1. Carefully put the 4081 IC into your breadboard with pin 1 near the negative bus. Reversing the IC usually destroys it. One power lead of an IC must be connected to a positive bus and the other to a negative bus. Connect a wire to one of the holes in the column associated with pin 7 and attach the other end to the negative bus. Likewise, connect a wire from a hole associated with pin 14 to the positive bus. One of the AND gates is connected to pins 1, 2, and 3. Connect the free end of the test probe to the AND gate output, pin 3. Connect one long wire to a hole in the column that corresponds to pin 1 and another long wire to a hole in the column that corresponds to pin 2. These two wires will be connected to the negative or positive bus to provide the input logical 0 and 1.

29-2 Physics Lab

4. Connect the 9-V battery. Let the wire connected to pin 1 be input *A* and the wire connected to pin 2 be input *B*. Record your observations in Table 1 under the output column *Z*. Connect both inputs to the negative bus, producing a 0,0 input to the AND gate. Observe the test probe LED. Move input *B* to the positive bus, producing a 0,1 input to the AND gate. Observe the test probe LED. Move input *A* to the positive bus and input *B* to the negative bus, producing a 1,0 input to the AND gate. Observe the test probe LED. Finally, connect input B to the positive bus, producing a 1,1 input to the AND gate. Observe the test probe LED. Demonstrate the AND gate logic to your teacher before proceeding.

Figure C

OR gate symbol   4071 IC

5. Figure C shows an OR gate symbol and is mathematically represented by $Z = A+B$, which means *Z* equals *A* or *B*. The figure shows a schematic representation of a 4071 IC containing four OR gates (quad 2-In OR gate). The electric connections are the same as those for the 4081. Without removing the wire connections, replace the 4081 IC with the 4071. Repeat the steps to test the four possible input combinations and record your output observations in Table 2. Demonstrate the OR gate to your teacher before proceeding.

Figure D

NOR gate symbol   4001 IC

6. Figure D shows a NOR gate symbol and a schematic representation of a 4001 IC containing four NOR gates (quad 2-In NOR gate). The NOR is mathematically represented by $Z = \overline{A+B}$, which means *Z* equals not *A* or not *B*. The electric connections are the same as those of the 4071. Without removing the wire connections, replace the 4071 IC with the 4001 IC. Repeat the steps to test the four possible input combinations and record your output observations in Table 3. Demonstrate the NOR gate to your teacher before proceeding.

29-2 Physics Lab

Figure E

NAND gate symbol

4011 IC

7. Figure E shows a NAND gate symbol and a schematic representation of a 4011 IC containing four NAND gates (quad 2-In NAND gate). The NAND is mathematically represented by $Z = \overline{A \cdot B}$, which means Z equals not A and not B. The electric connections are the same as those of the 4001. Without removing the wire connections, replace the 4001 IC with the 4011 IC. Repeat the steps to test the four possible input combinations and record your output observations in Table 4. Demonstrate the NAND gate to your teacher before proceeding.

8. Gates are usually combined to produce other logic combinations or to produce a simple logic output when the desired logic gate is unavailable. Use your 4011 quad NAND gate IC and wire the circuit shown in Figure F. Remember to wire the power to the IC through pins 7 and 14 as before. Test the possible combinations for inputs A and B. Record your results in Table 5 for the combination of NAND gates. Demonstrate the NAND gate combination to your teacher.

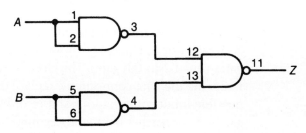

Figure F

Data and Observations

Table 1		
AND Gate Truth Table		
Inputs		Output
A	B	Z
0	0	
0	1	
1	0	
1	1	

Teacher's signature _____

Table 2		
OR Gate Truth Table		
Inputs		Output
A	B	Z
0	0	
0	1	
1	0	
1	1	

Teacher's signature _____

Table 3		
NOR Gate Truth Table		
Inputs		Output
A	B	Z
0	0	
0	1	
1	0	
1	1	

Teacher's signature _____

29-2 Physics Lab

Table 4
NAND Gate Truth Table

Inputs		Output
A	*B*	*Z*
0	0	
0	1	
1	0	
1	1	

Table 5
NAND Gate Combination Truth Table

Inputs		Output
A	*B*	*Z*
0	0	
0	1	
1	0	
1	1	

_____
Teacher's signature

_____
Teacher's signature

Analysis and Conclusions

1. Compare the truth table results for the combination of NAND gates (Table 5) with the other circuits and determine the type of logic it performs.

_____

_____

2. Compare the OR and NOR functions.

_____

_____

3. Compare the AND and NAND functions.

_____

_____

4. Predict the result of hooking together the inputs on a NAND gate.

_____

_____

5. Predict the result of hooking together the inputs on a NOR gate.

_____

_____

29-2 Physics Lab

Extension and Application

1. Digital ICs are found in common electronic equipment, such as automobiles, cellular phones, dishwashers, stoves, security systems, and telephones. Simple logic gates can do simple functions, such as turning on an alarm. In a car, for example, if the ignition is turned on and the gearshift is engaged, and if either of the seats is occupied and the corresponding seat belt is not fastened, an alarm will sound. Use the symbols below to write an appropriate logic statement for an alarm, using "+" for an OR and "·" for an AND.

Symbol	Information
A	alarm on
I	ignition on
L	left front seat occupied
B_L	left front seat belt not fastened
R	right front seat occupied
B_R	right front seat belt not fastened
G	gearshift engaged

2. Use the 4011 NAND gate IC to wire the circuit shown in Figure G. This circuit is useful in many applications and is called an Exclusive-OR (XOR). Write a truth table for the various input combinations.

Figure G

Four NAND gates combined to produce an XOR

30-1

Physics Lab

Radiation Properties

The three common types of radiation emitted by radioactive substances are alpha, beta, and gamma radiation. An alpha particle is a helium nucleus, a beta particle is an electron, and a gamma ray is a high-energy photon. Radiation can be detected by a Geiger-Mueller tube. The tube contains a gas that ionizes easily. Two electrodes are in the tube; one is a metal cylinder, and the other is a wire along the tube's center. A thin window allows radiation to enter the tube. A voltage is applied to the tube and can be adjusted to a value just below the point where spontaneous ionization would take place. When a radioactive particle enters the tube, it ionizes a gas atom. The electrodes attract the gas ion and the electron, and as they move through the tube, they collide with other gas atoms and ionize them. This avalanche of ions creates a current pulse through the tube, and this current generates a voltage across an external resistor. The voltage is amplified to produce an audible signal, deflect a meter, or advance a counter.

Objectives

- **Measure** radioactivity with a Geiger counter.
- **Analyze** the relationship between radiation intensity and distance to a radiation source.
- **Infer** the inverse-square law for radiation intensity.

Materials

Geiger counter

beta source

tweezers

gloves

stopwatch

meterstick

masking tape

graph paper

Procedure

CAUTION: *Do not bring any food, drink, or makeup into the laboratory. Handle the radioactive materials with tweezers, tongs, or gloves. Wash your hands with soap and water before you leave the laboratory.*

1. Set up the Geiger counter according to your teacher's instructions. Handle the Geiger-Mueller tube carefully, as it is fragile and expensive. Carefully lay the tube on a lab table and secure it with a piece of masking tape, as shown in Figure A. Place the meterstick with one end next to the Geiger-Mueller tube window.

Masking tape

Radioactive source

Geiger-Mueller tube

Meterstick

Figure A

2. Set the radioactive samples about 1–2 m from the Geiger-Mueller tube so that their influence on the detector is minimal. Turn on the counter and measure the background radiation for 1 min. On the line above Table 1, record the background radiation in counts per minute (c/min).

3. Place the beta source 2 cm from the window of the Geiger-Mueller tube. The sensitivity of Geiger counters varies, as does the activity of the radioactive samples, which depends on age and type. If the sample seems to be too active, move it several centimeters farther away from the tube. Turn on the counter and measure the activity for 1 min. Record in Table 1 the measured activity value and the distance of the source from the tube.

30-1 Physics Lab

Name _____

4. Move the beta source 1 cm farther from the tube and repeat the measurement. Record in Table 1 the measured value of activity at that distance. Move the radioactive source and take measurements for 1-min. periods until the activity drops to a value equal to or lower than the background radiation level. Record the measured activity in the table.

Data and Observations

Background radiation = _____ c/min

Table 1				
Distance (cm)	Measured activity (c/min)	Adjusted activity (c/min)	Distance^2 (cm^2)	1/Distance^2 (cm^{-2})
1				
2				
3				
4				
5				
6				
7				
8				
9				
10				
11				
12				
13				
14				
15				
16				
17				
18				
19				
20				

30-1 Physics Lab

Analysis and Conclusions

1. Calculate the adjusted activity for each reading by subtracting the background activity from each of the measured values; record the adjusted values in Table 1. Find the squares of the distances and record these values. Calculate and record the reciprocals of the distances squared.

2. Plot a graph of activity versus distance, with adjusted beta activity on the y-axis and distance on the x-axis.

3. Plot a graph of measured activity versus the reciprocal of distance squared, with the activity on the y-axis and the reciprocal of distance squared on the x-axis.

4. What is the source of background radiation that you measured?

_____

_____

_____

_____

_____

_____

5. What do your two graphs reveal about the nature of radiation and the relationship of radioactive intensity with distance?

_____

_____

_____

_____

_____

_____

_____

6. Was it necessary to subtract the background radiation? Explain.

_____

_____

_____

_____

_____

_____

30-1 Physics Lab

Extension and Application

1. Repeat the experiment with the gamma source. Set up a data table to record the measured activity for each distance from the source. Do the necessary calculations and plot the two graphs, as described in Analysis and Conclusions steps 1–3.

2. A radioactive source has a measured activity of 10 000 c/min at a distance of 5 cm. What activity would be expected at three times this distance?

_____

_____

_____

_____

31-1

Physics Lab

Radioactive Shielding

Objectives

- **Demonstrate** the ability of different types of radiation to penetrate materials.
- **Compare** the shielding efficiencies of various materials.
- **Communicate** how to determine what type of radiation is given off by a radiation source.
- **Infer** the relative shielding effectiveness of various materials.

Alpha, beta, or gamma radiation is released when the nucleus of an atom changes. Recall that an alpha particle is a helium nucleus, a beta particle is an electron, and a gamma ray is a high-energy photon. The mechanics of radiation absorption vary with the type of radioactive source, the initial energy of the radioactive particle or ray, and the type of absorbing material. The initial energy of the alpha particle is closely related to its penetration of a particular absorbing material. Because the alpha particle is a helium nucleus with a double-positive charge, however, its state of ionization makes it extremely interactive. Thus, despite its relative heaviness, an alpha particle is quickly absorbed. Since beta particles have the same mass as the electrons in the absorber, the beta particle is deflected in collisions with other electrons; it does not follow a well-defined path through the material. For beta particles, penetration is inversely proportional to the density of the absorbing material.

While charged particles gradually lose their energy in many collisions, photons lose all their energy in a single collision. Therefore, absorption of gamma rays is defined in terms of the absorption coefficient, the reciprocal of which yields the thickness of the absorber that reduces the number of photons in a beam by a certain percentage. Thus, the intensity of a gamma ray decreases exponentially as it penetrates a given material.

Procedure

CAUTION: *Do not bring any food, drink, or makeup into the laboratory. Handle the radioactive materials with tweezers, tongs, or gloves. Wash your hands with soap and water before you leave the laboratory.*

Materials

Geiger counter or scaler

stand for Geiger-Mueller tube

alpha, beta, and gamma sources

5-cm squares of thin cardboard, aluminum, and lead (10 of each)

tweezers, tongs, or gloves

stopwatch or electronic counter

1. Set up the Geiger counter, according to the teacher's instructions. Handle the Geiger-Mueller tube carefully, as it is fragile and expensive. Place the Geiger-Mueller tube in a stand so that it is vertical, as shown in Figure A.

Figure A

Geiger-Mueller tube

Radioactive source

31-1 Physics Lab

2. To obtain many readings in the short time available, you will make counts for only 10 s. Since radioactive decay is spontaneous and does not always occur at a constant rate, it is possible that some of the measured activities may be a little larger or smaller than expected. Several types of Geiger counters have electronic timers that turn on the counter for specific periods of time. If your Geiger counter has an electronic timer, the teacher will describe how to use it. Otherwise, use a stopwatch to time the counting. Turn on the counter and measure the background radiation for 10 s. Record this value on the line above Table 1.

3. Make sure that the radiation sources you are not measuring are far from the tube, or large errors could result. Place the alpha sample under the tube. Most alpha sources are fairly weak, so place the alpha source close to the window of the tube. Measure the activity for 10 s and record the value in the table. Place one sheet of cardboard on top of the alpha source. Measure the activity for 10 s and record the value in the table. Add a second sheet of cardboard and measure and record the activity. Add layers of cardboard in pairs until the activity drops to the background radiation level. Record all values in the table.

4. Remove the cardboard and repeat Step 3 with aluminum sheets. Once the measured activity drops to the background radiation level, stop measuring.

5. Remove the alpha source and repeat steps 3 and 4 with the beta source, testing the shielding capacity of cardboard, aluminum, and lead.

6. Replace the beta source with the gamma source. Repeat steps 3 and 4 with layers of cardboard, aluminum, and lead.

7. Return the radioactive sources to their appropriate storage containers, as instructed by the teacher.

Data and Observations

Background radiation = _____ c/10 s

Number of sheets	Cardboard alpha	Cardboard beta	Cardboard gamma	Aluminum alpha	Aluminum beta	Aluminum gamma	Lead beta	Lead gamma
0								
1								
2								
4								
6								
8								
10								

Table 1 — Measured Activity (c/10 s)

31-1 Physics Lab

Analysis and Conclusions

1. Which type of radiation is most easily absorbed? Which type of radiation is least easily absorbed?

_____

_____

2. How much and what type of material is required to reduce the various radiation sources to one half of their activity?

_____

_____

3. Explain how to identify an unknown radiation source.

_____

_____

_____

_____

4. Was it possible in this experiment to eliminate all of the gamma radiation? Give a reason for your answer.

_____

_____

Extension and Application

1. How thick must an absorber be to reduce to zero the intensity of a photon beam?

_____

2. One method for maintaining a consistent thickness of paper during its manufacture is to continually measure the intensity of radiation that passes through the paper and adjust the machinery as necessary. Which type of radiation would be most suitable for this purpose and why?

_____

_____

_____

_____

_____

1-1

Bubble Up

Introduction

Soaking a raisin increases its surface area and volume, so that its density is close to the density of water. The bubbles of carbon dioxide in carbonated water stick to the surface of the raisin; this raisin-bubble system is less dense than water, so that it floats. When the raisin-bubble system loses bubbles at the surface, it gets denser again and falls. As the water loses carbonation, the amount of time between flotations increases. This experiment is a good demonstration of indirect relationship between variables.

Process Skills

observing and inferring; measuring in SI; hypothesizing; experimenting; using variables, constants, and controls; interpreting data; communicating

Preparation Hints

- Soak the raisins overnight, or about 12–15 h, so that their density is nearly that of water.

Materials

100-mL beaker
150-mL beaker
100 mL carbonated water
4 raisins
clock
100 mL tap water
plastic spoon
metric ruler

Procedure Hints

- Have students work in teams of two to encourage cooperation, collaboration, and communication.

- You could let students try a number of approaches, even if they are totally unrelated (e.g., vary the intensity of light used to make observations). Two conditions are necessary to move the raisin: soaking and carbonation. To prove this, students will have to manipulate one condition (variable) at a time. You may prefer to guide students in this direction for their investigation of the cause.

- Students could test popular brands of carbonated water or soft drinks to compare raisin motion in different liquids.

- Do not explain why the raisins move up and down. Leave the question open. As with many scientific phenomena, a complete explanation may not be known. Instead, lead a discussion on which hypothesis seems most correct and why.

- Encourage students to do the activity at home to involve their parents.

Data and Observations

Table 1 (Sample data are given.)	
Material	**Observations**
Carbonated water	Clear and colorless; bubbles rising; bubbles coming out of solution; bubbles on bottom of beaker
Soaked raisin	Brown; plumped up; still has wrinkles; mushy; 14 mm–18 mm long; 9 mm–12 mm wide
Dry raisin	Slightly smaller, more wrinkled and oblong shaped, and firm compared to soaked raisins; 12–14 mm long; 8–12 mm in wide

Table 2 (Sample data are given.)
Observations of soaked raisin in carbonated water
Small bubbles will accumulate on the surface of the soaked raisin. Some of the bubbles gradually grow in size. The raisins roll around on the bottom of the beaker. After a short time, the raisins rise to the surface. Bubbles are released from the raisin where it surfaces, and the raisin then sinks. The raisin continues to sink and surface every 15–25 s, depending on the amount of carbonation in the water.

Analysis and Conclusions

1. observations of the raisin's appearance
2. observations of bubbles attaching to and coming off the raisin and of the rate of sinking and rising

3. Students may have a variety of hypotheses, some of which may not relate to their data. Sample hypothesis: The bubbles (carbon dioxide) come from solution and accumulate on the raisin until the density of the raisin has been reduced enough to cause it to float. The raisin then rises to the surface of the liquid.

4. Students should develop an experiment to determine if the carbonated water causes the raisin to rise and sink. The tap water can be a variable in this investigation. Students should also consider the soaking of the raisin and its effect. This variable can be checked by using a dry raisin, which in carbonated water will sink to the bottom and stay there. Students should manipulate only one variable at a time in their experiments. For example, the effect of tap water, carbonated water, or a mixture of the two should be varied while using only soaked raisins. Next, students can manipulate the type of raisin while keeping the type of liquid constant.

5. Answers will vary. Some students will find that their results did not support their original hypothesis. You might encourage them to state a new hypothesis based on their observations.

Extension and Application

1. The overall density of most people is slightly less than that of water, so they float. When people exhale, they become more dense, and displace less water, and sink. Each person has a unique body structure and composition, with varying amounts of bone mass, fat, and muscle. This individuality accounts for differences in body density and in the ability to stay above water. Those who are just barely able to float will readily sink after they exhale.

2. Air pumped into a sunken ship makes it less dense, so that it can float.

3. A submarine has ballast chambers filled with either water or air. To sink, the submarine replaces air in the chambers with water. To rise, it replaces water with air.

2–1
Measuring Length

Introduction

Measurement errors result from parallax and from incorrect use of significant digits. Precision improves when students make multiple readings and calculate the average measurement.

Misconception: Students may misuse the words *precision* and *accuracy*. *Precision* is the uncertainty of a measurement. If many measurements of a sample deviate very little from the average, the measurement is precise. *Accuracy* means that the measurement conforms to a standard measurement. Calibration of an instrument assures its accuracy.

Process Skills

measuring in SI; using numbers, observing and inferring

Preparation Hints

- A large object such as a table or even the dimensions of the classroom can be substituted for either the wood block or the Laboratory Manual.

Materials

meterstick
wood block
Laboratory Manual

Procedure Hints

- Select several identical objects, such as chalkboard erasers, textbooks, and pieces of chalk, and give them to different groups to measure. Compare each group's results to stimulate a discussion on precision and accuracy. Emphasize that researchers must develop a certain level of confidence in the data derived from experimentation.

- Caution students about significant digits. Help them understand that when they perform any mathematical operation with measurements, the end product cannot be more precise (or have more significant digits) than the initial measurements.

Data and Observations

Table 1 (Sample data are given.)

	Length (cm)	Width (cm)	Height (cm)	Volume (cm³)
Wood block	11.04	3.00	3.74	124
Laboratory Manual	27.70	22.80	1.25	789

Table 2 (Sample data are given.)

Number of pages	Total thickness (mm)	Average thickness of a single page (mm)
1	0.1	–
50	4.5	0.09
75	7.0	0.093
100	9.0	0.09

Analysis and Conclusions

1. The ends may be worn or damaged, so that increments are missing.

2. The estimate of one page; the uncertainty is large when measuring the thickness of a single piece of paper.

3. The most accurate value will be the one derived from the measurement of 100 pages. The uncertainty in measuring is the same for all of the measurements, so the more pages in the sample, the less error per page.

Extension and Application

1. A picture frame won't look good or work well unless the corners are square. If the measuring instrument does not produce precise angles, then the frame pieces will not align properly, even if they are cut and assembled by a skillful craftsperson.

2-2
Measuring Temperature

Introduction

Students will devise a procedure for measuring and comparing the temperatures of three samples of heated water drawn from the same source. They will read temperatures at various times for each sample and will graph the data to interpret their results. Sample volume will affect the reliability of temperature readings. In this lab, students will apply the concepts of measurement, accuracy, and precision discussed in Chapter 2 of their textbook.

Process Skills

hypothesizing; using variables, constants, and controls; measuring; comparing and contrasting; collecting and organizing data; making and using graphs

Possible Hypotheses

The correct hypothesis is that smaller sample volumes produce lower temperature readings and that the measurement process affects the outcome. Students may incorrectly hypothesize instead that the volumes are irrelevant or that measurement itself has no effect on the results.

Possible Materials

CBL unit
temperature probe
graphing calculator with PHYSICS CBL software
    program
link cable
beaker, 250-mL
3 beakers, 150-mL
3 10-mL graduated cylinders
100-mL graduated cylinder
tap water
4 thermometers, 210°C to 100°C range
safety goggles
heat-protective gloves
tongs, beaker
electric hot plate
stopwatch or watch with second hand

Possible Procedures

The following procedure works well. Use a hot plate to heat 250 mL of water in a beaker to about 65°C. Pour 80 mL of the heated water into a 100-mL graduated cylinder, 5 mL into a 10-mL

graduated cylinder, and 1 mL into a 10-mL graduated cylinder. Read the three temperatures and
record them every 30 s for 3 min. The CBL can
take temperature readings each 0.03 s for about
3 minutes.

Students might use a different heat source, such
as a Bunsen burner, to warm the water. They might
take the samples at separate times, and use a single
thermometer to take all the readings. They might
use other glassware to hold the samples.

Teaching Suggestions

- Student groups of three or four can simultaneously gather data in a short time. Each
  student should collect some of the data.
  Students should exercise caution using the
  heat source and handling the heated container of water. They should use tongs or
  heat-protective gloves and wear goggles.

- Review students' plans, and accept reasonable procedures. Also review graphing techniques, and make sure that the concepts
  involved are clear.

Data and Observations

(Sample data are given.)

Room temperature: ____22°C____

Temperature of initial large sample of heated
water: ____65°C____

Table 1 Temperature Data			
Time (min)	80 mL	5 mL	1 mL
0.5	54	44	42
1.0	54	43	41
1.5	53	42	40
2.0	53	42	39
2.5	53	41	38
3.0	52	41	37

Analyze and Conclude

1. The plots should resemble the graph below.
   The slopes differ for the different samples:
   the smaller the sample, the steeper the
   slope. Temperatures are lower overall for the
   smaller samples.

2. The smaller volume sample lost more heat
   to the environment. Moving the heated
   water to other containers and putting thermometers in it affected the temperature
   readings. Time allowed to elapse between
   pouring the sample and beginning the readings and the placement of the bulb of the
   thermometer in the sample may also affect
   the measured temperature. The CBL temperature probe has less mass and affects the
   system less, so its results may be closer in
   temperature.

3. Answers will vary. The temperatures aren't
   as high as the original water source. You
   wouldn't expect them to be the same
   because they lose heat to the sample containers and thermometers, which are room
   temperature.

4. Heat would be lost to the surroundings
   until the sample temperatures matched the
   room temperature.

5. The water temperature in the tube would be
   higher than the ocean temperature; the
   tube, thermometer, and surrounding air
   would warm it up. After 5 min, the difference would be still greater.

6. Observation or measurement involves interaction with whatever is being measured. In
   order to assure accuracy, an experiment
   must be carefully planned ahead.

Apply

1. The cars are very massive and the energy of
   the striking radio waves is so small that any
   resulting change in the cars' speeds is negligible, so there is no important effect on the
   reliability of the values obtained.

3-1

Analyzing Motion

Introduction

If an object travels equal distances in equal time periods, it has a constant average velocity and a constant average speed. When the constant-velocity vehicle starts, its velocity increases and so it is in nonuniform motion for an initial period of time. The motion of the vehicle can be analyzed by examining its displacement or velocity per unit time.

Process Skills

classifying, defining operationally, using numbers, interpreting data

Preparation Hints

- An alternative to the use of a CBL unit is a standard tape-recording timer.

- Science Kit offers an economical timer that operates from the standard 120 V, 60 Hertz AC. It always vibrates at the line frequency of 60 Hertz so that the period of the timer is known to be 1/60 of a second. A more expensive spark timer that has switchable 60 Hertz and 10 Hertz spark rates is available from several other suppliers. The spark produces consistent, excellent results, but requires more expensive electrostatic recording paper. However, these types of timers eliminate the need for dry cells or power supplies.

- The recording timer can be replaced by an instant-developing camera, using ASA 3000 film, triggered by a shutter-release cable. Install an electronic blinky (a flashing light set to flash at regular intervals) on the constant-velocity vehicle. An air-track car could be used instead. Place a photographic meterstick behind the vehicle. Put the camera on a tripod and, in a darkened room, open the shutter with the cable. Turn on the blinky and the vehicle and allow the vehicle to move across the field of view. Turn on the room lights for about 5 s to expose the meterstick, and then close the camera lens by releasing the shutter cable. Develop the photograph. A series of bright dots (the blinky) should appear across the photograph. The moving vehicle will not be visible. The dots can be measured with the photographic meterstick.

Materials

constant-velocity vehicle
metric ruler
C-clamp
CBL unit
graphing calculator with PHYSICS CBL software program
link cable
ultrasonic motion detector
graph paper

Procedure Hints

- If the motion detector seems to produce erratic results, try moving it a little farther away from the moving vehicle. Additionally, to make the vehicle appear more easily to the motion detector, a 5 × 7-inch index card can be taped to the vehicle.

Data and Observations

Table 1 (Sample data are given.)			
Time (s)	Displace- ment (m)	Total displace- ment (m)	Average velocity (m/s)
0.9	0.45	0.45	—
1.2	0.064	0.514	0.21
1.5	0.049	0.563	0.19
1.8	0.054	0.617	0.19
2.1	0.105	0.722	0.23
2.4	0.065	0.787	0.22
2.7	0.008	0.795	0.19
3.0	0.141	0.936	0.23
3.3	0.032	0.968	0.22
3.6	0.072	1.04	0.22

Analysis and Conclusions

1. Answers will vary. Students should describe the motion of the vehicle as moving slowly at the start, then moving faster until the speed appears to become constant.

2. The displacements are small at the beginning of the tape. They increase during the next few intervals and then become constant.

3. Basically constant and approximately zero.

4. Yes, the first few intervals show a change in velocity because the vehicle is accelerating.

5. Yes, most of the time intervals after the first few appear to be constant because the vehicle is not accelerating.

6. Answers for the vehicle's average velocity for the entire trip will vary. Because the period of changing velocity is so brief, the average velocity will likely be close to the value of the uniform velocity.

Extension and Application

1. For the sample data, the average velocity is $0.004 \text{ m/s}^2 \approx 0$.

2. a. The distance would double.

    b. The overall displacement would double.

    c. The acceleration will still be approximately zero. It is moving at a constant speed.

4–1

Addition of Vector Forces

Introduction

Vectors can be used to represent forces that act in combination. Forces applied at the same time to an object can balance one another, so that the object is in equilibrium and there is no motion. Various methods can be used to measure the forces acting simultaneously. The laws of vector addition can then be used to find resultant forces and to compare the resultants of various combinations of the forces. In this activity, vector properties and components will be explored, as well as the process of determining magnitudes and directions of vector sums.

Process Skills

observing, using numbers, formulating models, interpreting data, measuring in SI, predicting, defining operationally

Materials

Method 1—force table and 3 spring scales
Method 2—2 spring scales, 2 ring stands, cross support with rings or hooks, and 500-g hooked mass
metric ruler
heavy string
pencil
protractor
paper, plain white

Preparation Hints

- The activity and data collection are easily and quickly performed. Even with a limited number of equipment setups, all students should be able to obtain data in a class period if they have previously read the experiment.

Procedure Hints

- Make sure students check that the scales register zero at no load. Explain how to make zero corrections if the scales need adjustment.

- Remind students that the lengths of the vectors are proportional to the forces, not to the lengths of the strings or the spring scales.

- After completing the experiment, students should realize that the order of vector addition does not affect the final result (commutative law). Also, the grouping of vectors during addition does not affect the final result (associative law).

Data and Observations

Drawings should be similar to those in Figure B.

Analysis and Conclusions

1. Answers will vary.

2. The resultant of *A* + *B* should be nearly equal in magnitude to and opposite in direction from *C*. Relative error should be less than 15%.

3. Diagrams will vary.

4. The sum of *A* + *B* + *C* is a closed triangle indicating equilibrium. (No net force is exerted on point P.)

5. If *B* is added to *C*, the resultant should be equal in magnitude to and opposite in direction from *A*.

6. If *C* is added to *A*, the resultant should be equal in magnitude to and opposite in direction from *B*.

7. *C* + *B* + *A* should give the same result as *B* + *C* + *A*. The resultant is zero.

Extension and Application

1. The results should be the same as those obtained earlier. The sum of $A + B + C$ is zero.

2. When a body falls through a fluid such as air, a frictional or drag force acts on it, slowing its downward velocity. As it begins to fall, its velocity increases, but so does the frictional force acting on it. When the upward frictional force equals the downward force due to gravity, the net force acting on the body is zero. At this point, equilibrium is reached, and the body falls at its constant terminal velocity. The frictional force of the air that acts on the skydiver's body, or the body and the parachute, depends on the configuration of the body. When the total surface area of the falling body exposed to the frictional force increases, the terminal velocity decreases.

5–1

Accelerated Motion

Introduction

This lab uses graphs to demonstrate relationships between time and velocity, displacement, and acceleration. As an object undergoes constant acceleration, its velocity over time changes—made clear by the fact that the graph of velocity versus time is a straight line with a constant slope. The graph of displacement versus time is a parabola with constantly changing slope. The graph of acceleration versus time is a horizontal line.

Process Skills

measuring in SI, collecting and organizing data, interpreting data, making and using graphs, using numbers, analyzing, comparing and contrasting

Preparation Hints

- You may need to tape a 5 × 8 index card to the back of your cart so it more visible by reflecting sound waves to the motion detector

- Remind students that displacement is measured from the motion detector.

Materials

laboratory cart
CBL unit
link cable
graphing calculator with PHYSICS CBL software program
500-g mass
masking tape
heavy string, 1 m
pulley, double mounted
1 C-clamp
graph paper

Procedure Hints

- Students should release the cart gently so as not to have it start with a jerk. Holding the string is another way to start the system.

- If the cart is accelerating quite slowly, you may wish to change the time interval from 0.02 to a larger number.

Data and Observations

Table 1 (Sample data are given.)	
Time (s)	Displacement (m)
1.0	0.42
1.08	0.453
1.16	0.51
1.24	0.578
1.32	0.667
1.40	0.773
1.48	0.896

Table 2 (Sample data are given.)	
Time (s)	Average velocity (m/s)
1.0	0
1.08	0.41
1.16	0.713
1.24	0.85
1.32	1.11
1.40	1.33
1.48	1.6

Analysis and Conclusions

1.

Total Displacement vs Time

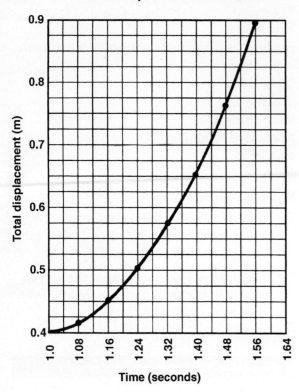

Time (seconds)

2.

Average Velocity vs Time

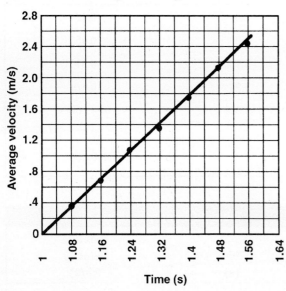

Time (s)

3. The graph is a parabola. The constantly increasing slope indicates that the cart was traveling greater distances during successive intervals of time.

4. The graph is a straight line passing through the origin. It indicates that the velocity of the cart increased at a constant rate.

5. Answers will vary. The unit for the slope is centimeters over the interval squared.

6. The graph is a straight line parallel to the *x*-axis. Acceleration is uniform.

Acceleration vs Time

Time (s)

Extension and Application

1. Answers will vary. The slopes should be close to the values for average velocity.

2. Answers will vary. The acceleration of the cart is less than the acceleration due to gravity because the falling mass is accelerating itself and the cart.

3. Passengers would experience the same acceleration that exists on Earth. Higher acceleration might cause physical discomfort. If the spacecraft could accelerate at that rate for a year, it would be traveling at close to the velocity of light, 3×10^8 m/s.

5-2

Acceleration Due to Gravity

Introduction

In this lab, students devise a procedure for measuring displacements and times for a freely falling object, most likely making use of a recording timer. They will use the resulting data to calculate velocity and velocity change, and to derive a value for the acceleration due to gravity. In carrying out these activities, they will apply the concepts of displacement, velocity, and acceleration discussed in Chapter 5 of their textbook.

Process Skills

hypothesizing, observing, measuring, interpreting data, making and using graphs, calculating

Possible Hypotheses

A recording timer can collect displacement and time information that can be used to calculate a value for g that is close to the accepted value, 9.80 m/s^2.

Possible Materials

recording timer with power supply
timer tape
carbon-paper disks
C-clamp
masking tape
1-kg mass
centimeter ruler

Possible Procedures

One procedure that works well is to use masking tape to attach the 1-kg mass to a timer tape in a recording timer. Start the timer and drop the mass. Turn off the timer, remove the tape, and analyze it. Use the distance between the dots and the period of the timer to calculate average velocity.

Some students may choose the procedure just described, perhaps varying some of the materials. Others may want to use a stopwatch to determine falling times. You might suggest to these students that they try these procedures before the actual laboratory session so that they can see that the rapid drop and the short distances make it very difficult to obtain accurate information this way.

Teaching Suggestions

- Students should know how to operate the recording timer if they've already done Labs 3.1 and 5.1. Use timers with a period between 0.016 and 0.020 s. Limit students to a fall height of between 0.5 and 0.6 m.

- Review students' plans, and accept reasonable procedures. It is important that students have a workable plan for deriving displacements and times. You might want to review calculation and graphing techniques, as well as the concepts involved.

- Inform students that they must not apply any upward or downward force when they release the mass.

Data and Observations

Actual displacements and times depend upon the period of the recording timer, but results should show relatively constant increases in time and displacement during the intervals, with average velocity increasing as well.

Period of recording timer: ___0.0167___ s

Analyze and Conclude

Table 1 Time, Displacement, and Velocity Data (Sample data are given.)				
Interval	Total time (s)	Displacement in interval (m)	Total displacement (m)	Average velocity (m/s)
1	0.017	0.0030	0.0030	0.18
2	0.033	0.0065	0.0095	0.39
3	0.050	0.0090	0.0185	0.54
4	0.067	0.0120	0.0305	0.72
5	0.083	0.0145	0.0450	0.87
6	0.100	0.0170	0.0620	1.02
7	0.117	0.0195	0.0815	1.17
8	0.133	0.0220	0.1035	1.32
9	0.150	0.0250	0.1285	1.50
10	0.167	0.0280	0.1565	1.68
11	0.183	0.0300	0.1865	1.80
12	0.200	0.0330	0.2195	1.98
13	0.217	0.0360	0.2555	2.16
14	0.233	0.0380	0.2935	2.28
15	0.250	0.0410	0.3345	2.46
16	0.267	0.0435	0.3780	2.61
17	0.283	0.0460	0.4240	2.76
18	0.300	0.0490	0.4730	2.94
19	0.317	0.0510	0.5240	3.06
20	0.333	0.0540	0.5780	3.24

1. See Table 1 for sample values.

2. If students graph their results, they should get a parabolic curve, as in the graph below. The constantly increasing slope shows that the mass was traveling greater and greater distances during successive time intervals, indicating acceleration.

3. If students graph their results, they should get a straight line through the origin with a positive slope, as in the graph below. The velocity of a falling mass increases at a constant rate indicating a constant acceleration.

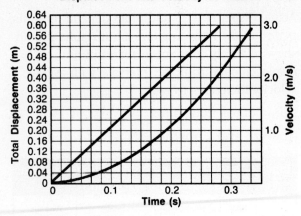

Displacement and Velocity vs Time

4. Sample calculation:

$$a = \frac{v_f - v_i}{t_f - t_i}$$

$$= \frac{3.24 \text{ m/s} - 0 \text{ m/s}}{0.333 \text{ s} - 0 \text{ s}}$$

$$= 9.73 \text{ m/s}^2$$

The values calculated by the students should be reasonably close to 9.80 m/s². The hypothesis that a recording timer can be used to find a value close to the accepted value should be borne out.

Apply

1. The value of g at very high elevations would be slightly lower. Athletes in jumping or throwing competitions might be helped because objects would remain in the air slightly longer, yielding better distances. The advantage would be very slight, however, and could well be offset by disadvantages such as lower oxygen levels at high altitude.

6–1

Newton's Second Law

Introduction

Newton's second law of motion indicates that an object will accelerate if, and only if, it has a net force acting on it. Further, the law indicates that the acceleration will be in the direction of the force.

Process Skills

observing, using numbers, measuring, communicating, interpreting data, questioning, controlling variables

Preparation Hints

- A large-capacity electronic balance (4 kg) will speed the process of determining the mass of the carts and attached tape and string.

- Use the following program, L03ACCL, for this activity. Enter in order:

```
PROGRAM:L03ACCL
:Lbl 0
:ClrHome
:Menu("*MAIN MENU*"),"RUN
ACTIVITY",A,"QUIT",Z)
:Lbl A
:PlotsOff
:Func
:AxesOn
:FnOff
:ClrList L2,L3
:Plot1(Scatter,L2,L3, · )
:0→Xmin
:10→Xmax
:.2→Xscl
:0→Ymin
:10→Ymax
:.5→Yscl
:60→dim L2
:60→dim L3
:seq(I,I,.1,6,.1)→L2
:{1,0}→L1
:Send(L1)
:{1,11,3}→L1
:Send(L1)
:ClrHome
:Disp "PRESS ENTER"
:Disp "TO START"
:Disp "ACCL GRAPH."
:Pause
:ClrDraw
:Text(4,1,"DIST")
:Text(51,78,"TIME")
:(3,.1,–1,0)→L1
:Send(L1)
:For(I,1,60,1)
:Get(L3(I))
:Pt–On(L2(I),L3(I))
```

```
:End
:Stop
:Goto 0
:Lbl Z
:ClrHome
```

Materials

laboratory cart
2 250-g masses
1000-g hooked mass
set of small masses
pulley, single-mounted
C-clamp
heavy string (1.5 m)
masking tape
balance
meterstick
CBL unit
vernier motion detector
TI-92 calculator

Procedure Hints

- Students should be aware that the time expressed using the motion detector may not correlate exactly with real time. They should ignore this possible discrepancy and use the time found by the motion detector.

Data and Observations

Table 1 (Sample data are given.)

	Value
Mass of laboratory cart (kg)	1.292
Mass needed to equalize friction of cart (kg)	0.015

Table 2 (Sample data are given for 0.5 m.)

Trial	Distance (m)	Accelerating force (N)	Time (s)
1	0.5	9.80	0.52
2	0.5	9.80	0.50
3	0.5	9.80	0.50

Table 3 (Sample data are given.)

Distance	Acceleration (m/s^2)	Total mass (kg)	$(m)(a)$ (N)
0.5 meter	3.70	2.308	8.53
1.0 meter	4.00	2.308	9.23
1.5 meter	4.00	2.308	9.23

Table 4 (Sample data are given.)

Trial	Distance	Acceleration (m/s^2)	Total mass (kg)	$(m)(a)$ (N)
empty cart	1.0 meter	4.00	2.308	9.23
cart + 250 g	1.0 meter	3.7	2.558	9.46
cart + 500 g	1.0 meter	3.4	2.808	9.55

Analysis and Conclusions

1. Trial 1: $a = 2(0.5 \text{ m})/(0.52 \text{ s})^2 = 3.70 \text{ m/s}^2$
   Trial 2: $a = 2(0.5 \text{ m})/(0.50 \text{ s})^2 = 4.0 \text{ m/s}^2$
   Trial 3: $a = 2(0.5 \text{ m})/(0.50 \text{ s})^2 = 4.0 \text{ m/s}^2$

2. Less than; the falling mass must move itself plus the cart. Therefore, the force of the falling mass is diminished by that amount of force required to move the cart's mass.

3. total mass = 1.293 kg (cart + string) + 1.000 kg (*falling mass*) + 0.015 kg (small masses) = 2.308 kg

4. Trial 1: $(2.308 \text{ kg})(3.70 \text{ m/s}^2) = 8.54 \text{ N}$
   Trial 2: $(2.308 \text{ kg})(4.00 \text{ m/s}^2) = 9.23 \text{ N}$
   Trial 3: $(2.308 \text{ kg})(4.00 \text{ m/s}^2) = 9.23 \text{ N}$

5. Error should be less than 20%, and with careful technique, it can be less than 10%. Using the sample data from the three trials, the values for relative error were 5%, 7%, and 4%.

6. $F = mg = (0.015 \text{ kg})(9.80 \text{ m/s}^2) = 0.15 \text{ N}$

7. Frictional forces oppose the force (weight) of the falling mass; the net force is thus reduced by this friction.

8. The mass of the cart should not have affected its acceleration.

Extension and Application

1. The frictional forces acting on an automobile include sliding friction between tires and the road and air resistance to automobile movement. When the frictional forces

are equal to the force produced by the engine, the net force is zero. Therefore, there will be no acceleration, and the automobile will travel at a constant velocity.

6-2

Friction

Introduction

In this lab, students will design a procedure for measuring and comparing the coefficients of static friction and sliding friction for an object on an inclined plane. They will gather data on weight, frictional force, and angle of tilt, and calculate the coefficients of friction. In carrying out this activity, they will apply the concepts of friction force and static and kinetic friction discussed in Chapter 6 of their textbook.

Process Skills

hypothesizing, measuring in SI, comparing and contrasting, making and using tables, interpreting data, using numbers

Possible Hypotheses

The correct hypothesis is that the angle for static friction is greater than that for sliding friction, and that the coefficient of static friction is greater than that of sliding friction. Students may incorrectly hypothesize that the angle and coefficient of sliding friction are greater than or equal to the angle and coefficient of static friction.

Possible Materials

CBL unit
link cable
graphing calculator with PHYSICS CBL software
    program
force sensor
DIN adapter
small rectangular object (such as a book, chalkboard eraser, or small block of wood)
wooden board
heavy string (1 m)
masking tape
protractor
meterstick

Possible Procedures

One procedure that works well is to tape a string to the selected object and use it to hang the object

from a spring scale to determine its weight. To determine the force of sliding friction, set a board flat on a table and, with the string parallel to the board, pull the object along at constant speed. Do a few trials and take the average.

Remove the string and spring scale from the object and place the object on the board. Lift one end of the board slowly. Record the tilt angle at which the object begins to slide. This value is the angle for static friction. Repeat the tilting process with the object set into slow motion by repeated taps to eliminate the effects of static friction. Measure the angle at which the object slides at constant speed. This value is the angle for sliding friction.

Instead of tapping the object, some students may suggest letting the object begin sliding at the angle of static friction, then quickly reducing the tilt to eliminate acceleration and produce motion at constant speed. This technique should be successful with some practice, assuming that the board is long enough.

Teaching Suggestions

- Have students work in groups of two or three. It would be very difficult for students working alone to manipulate the materials quickly enough to gather the required data. Each student should collect some of the data. You could have each student responsible for data for a different object.

- Allow students to practice before collecting data on sliding friction.

- The wood blocks used to store sets of masses are good experimental objects. Chalkboard erasers also work well.

- You could hint at a way to determine the coefficient of static friction without using a protractor. The tangent of the angle of tilt is the ratio of the height of the plane to its length. Thus, $h/l = \tan \theta = \mu$.

- The graphing calculator and force sensor will produce good results if students pull at a slow and steady rate so as to balance the frictional force.

- Please note the maximum range of the sensors students are using. Review the instructions that came with your sensor. The Student Force Sensor had an approximate maximum force of 10 N (1 kg mass block), while the Vernier Dual Range Force Sensor can go as high as 50 N.

Data and Observations

[Sample data are given.]

Kind of surface: wood

Object: chalkboard eraser

Weight of object: 1.86 N

Force of Sliding Friction	
Trial	Force, F_f (N)
1	0.40
2	0.38
3	0.39
Average	0.39

Angles of Slide		
Initial state	Angle, θ	Coefficient, μ
static	20°	0.36
sliding	13°	0.23

Analyze and Conclude

1. $F_\perp = F_w$, so $F_\parallel/F_\perp = F_\parallel/F_w = 0.39/1.86 = 0.21$

   $\mu_{sliding} = \tan 13° = 0.23$

   The two values should be equal or nearly so, given experimental error.

2. $\mu_{static} = \tan 20° = 0.36$

   The coefficient of static friction is larger than the coefficient of sliding friction.

3. The angles at which the brick begins sliding should be the same, regardless of the area of the surfaces in contact. The weight of the brick perpendicular to the surface influences the force of friction; the coefficient of static friction is independent of the area of contact.

4. An unwrapped brick begins to slide at an angle greater than that required by a brick wrapped in wax paper. The wax paper acts like a lubricant to reduce the coefficient of static friction and thus the force of friction.

5. The type of surface and the force normal to the surface (perpendicular component of weight) influence the force of friction.

Apply

1. Brand Y will give better service because it has greater traction in wet weather than brand X does. Both tires have nearly the same coefficient of friction on dry pavement and thus nearly the same amount of traction in dry weather.

6-3

Pushes, Pulls, and Vectors

Introduction

All forces can be analyzed as vector quantities. If a force is applied to an object at any angle other than parallel to the direction of motion, then the net force acting on the object in the direction of motion must be determined using vectors. As the angle through which a force is applied to an object increases, the net force acting on the object in the direction of motion decreases. If an object moves at a constant rate, then the force that is causing the motion is also constant. An electronic force sensor is used in this experiment to measure the force necessary to move an object at a constant rate. By measuring the applied force at several different angles and graphing the results, the students will be able to determine the net force acting to move the object. Because the same object is used for all trials, the calculated net force should be approximately the same in all cases. Discrepancies in the calculated values will be greatest at large angles because the upward component of the applied force tends to decrease the force of friction acting on the object and therefore less force is required to move the object.

Process Skills

measuring, making and using graphs, interpreting data, analyzing

Materials

meterstick
protractor
masking tape
book or other light- to medium-weight object
heavy string
CBL unit
vernier student force sensor with
    CBL DIN adapter
TI-82 calculator
TI-92 calculator

Preparation Hints

- Two class periods will be needed for this activity if entering the L07FORCE program.

- Be sure there is a large open area for students to work in.

- Use the following program, L07FORCE, for this activity. Enter in order:

```
PROGRAM:L07FORCE
:Lbl 0
:ClrHome
:Menu("*MAIN MENU*"),"RUN
ACTIVITY",A,"QUIT",Z)
:Lbl A
:PlotsOff
:FnOff
:AxesOn
:0→Xmin
:99→Xmax
:–5→Ymin
:10→Ymax
:10→Xscl
:2→Yscl
:ClrList L2,L4
:{1,0}→L1
:Send(L1)
:ClrHome
:{4,1,1,1,9.8,–9.8}→L1
:Send(L1)
:{1,1,1,0,0,1}→L1
:Send(L1)
:Disp "PRESS ENTER"
:Disp "TO START"
:Disp "FORCE GRAPH."
:Pause
:{3,.1,-1,0}→L1
:Send(L1)
:99→dim L4
:ClrDraw
:Text(4,1,"FORCE(N)")
```

```
:Text(51,81,"T(S)")
:For(I,1,99,1)
:Get(L4(I))
:Pt–On(I,L4(I))
:End
:seq(N,N,0,9.8,.1)→L2
:9.8→Xmax
:1→Xscl
:Plot1(Scatter,L2,L4, · )
:DispGraph
:Text(4,1,"FORCE(N)")
:Text(51,81,"T(S)")
:Stop
:Goto 0
:Lbl Z
:ClrHome
```

Procedure Hints

- Be sure all groups of students pull the same type and weight object. If the object is much lighter or heavier than a book, the window of the TI-82 will have to be adjusted.

- Have students work in groups of three or four.

- You may have students measure the pulling angle by using a protractor or by taking measurements with the meterstick and using trigonometry to compute the proper angle.

Data and Observations

Table 1 (Sample data are given.)		
	Amount of force applied (y-value)	Amount of horizontal force applied
45°	4.68	3.31
30°	4.44	3.86
15°	4.19	4.03
0°	3.95	3.95

Analysis and Conclusions

1. The horizontal force would be the same. Only the direction of the vertical force is changed.

2. Students should see that as pulls are closer to 0°, the force applied is less and is closer to the horizontal force applied.

3. Answers will vary. Students should extrapolate the data to 60°.

4. The force would not be one-half the value. A force applied at 90° would pull an object straight up, and would not produce any horizontal force. Students may use data from this activity to justify this answer.

Extension and Application

1. It is probably more difficult to move a small wagon by bending over to pull it (with a 0° pull) than it would be to move it by pulling it at waist (approximately 45°) level.

7-1

Projectile Motion

Introduction

The laboratory setup simulates projectile motion. It allows students to see that the path of the projectile is half of a parabola. The graph of the vertical distance covered by the steel ball is also a parabola, indicating that distance traveled is proportional to the square of the time. The graph of the vertical velocity shows that the steel ball was subjected to uniform acceleration. Students will be able to determine that the initial vertical velocity, v_y, and the initial horizontal velocity, v_x, are constant.

Process Skills

using numbers, observing, formulating models, communicating, interpreting data, measuring, questioning

Preparation Hints

If carbon paper is not available, the steel ball can be coated with tempera paint. A cafeteria tray works well for the incline. When rolled down the incline and across the paper, the steel ball will leave a wide trail, and the tray edge will usually stop the ball.

Materials

steel ball
1 sheet lined paper
masking tape
board or tray
grooved metric ruler
1 sheet carbon paper
textbook support
block support
TI-92 calculator
stopwatch or watch with second hand

Procedure Hints

- If students use narrow-ruled notebook paper, they should count every two or three lines as one interval of time. This will give them sufficient space in the table provided to record data from the entire width of the lined paper.

- Help students understand that the horizontal component of the ball's velocity is constant as the ball rolls across the inclined board. Because this is true, the lined spaces each represent equal time intervals. Although the time in seconds could be determined for each interval, it is not necessary to do so. Students can find the actual time interval using the equation time = total time ÷ number of tine intervals.

- Review with students how to input data into the TI-92 calculator. To simulate graph paper, students may want to set the window to x-min = -10, x-max = 10, y-min = -5, y-max = 5, x-scl = .1, and y-scl = 1.

Data and Observations

Table 1 (Sample data are given.)			
Interval number	Time (s)	Total vertical distance (m)	Average vertical velocity (cm/s)
0	0	0	0
1	0.1	0.35	0.035
2	0.2	0.80	0.045
3	0.3	1.25	0.055
4	0.4	1.80	0.055
5	0.5	2.35	0.055
6	0.6	3.00	0.065
7	0.7	3.70	0.070
8	0.8	4.45	0.075
9	0.9	5.25	0.080
10	1.0	6.00	0.075
11	1.1	6.80	0.080
12	1.2	7.65	0.085
13	1.3	8.55	0.090
14	1.4	9.50	0.095
15	1.5	10.60	0.11
16	1.6	11.50	0.09
17	1.7	12.60	0.11
18	1.8	13.80	0.12
19	1.9	14.90	0.11
20	2.0	16.20	0.13

Analysis and Conclusions

1.

Vertical Distance versus Time

2. The graph is a parabola, which indicates that the vertical distance traveled by the steel ball is proportional to the square of the time. The equation for the graph is that of a parabola in which v_y and g are constants:

$$y = v_y t + \frac{1}{2} g t^2$$

3. The path of the steel ball on the lined paper inscribed half of an arch. In math, this curved line is called a parabola.

4. Because v_x is a constant, students should describe their graphs as a straight line parallel to the y-axis. Its value represents the initial horizontal velocity.

5.

Vertical Velocity versus Time

6. A change in velocity over a period of time, $\Delta v/\Delta t$, is the definition of acceleration. If this change is uniform or constant, a straight-line graph results. Thus, as a projectile falls to Earth, it is subjected to the uniform acceleration due to gravity, g.

7. Because v_y is a constant, students should describe their graphs as a straight line parallel to the x-axis Its value represents the initial vertical velocity.

Extension and Application

1. The bullet will strike the tin can only if it falls off the shelf at the moment the bullet leaves the gun muzzle. If the can falls off the shelf, both it and the bullet will be accelerated downward by gravity at the same rate. The can will fall a vertical distance equal to $\frac{1}{2}gt^2$ from the shelf, and the bullet will have fallen the same vertical distance from its starting point when it reaches the line of fall of the target. When the bullet reaches the line of fall, it will be the same distance below the tin can's initial position on the shelf as the tin can, thus it will strike the can. The greater the initial velocity of the bullet, the higher above ground it will hit the tin can. If the tin can does not fall off the shelf, it will not be hit by the bullet.

7-2

Range of a Projectile

Introduction

In this lab, students will design a procedure to predict a projectile's horizontal range of motion. They will collect data on the horizontal component of velocity and on vertical distance of fall. They will also verify the concept that horizontal and vertical components of motion are independent. In carrying out this activity, they will investigate the concepts of projectile motion and motion in two directions, discussed in Chapter 7 of their textbook.

Process Skills

measuring in SI, using numbers, hypothesizing, interpreting data, making and using tables, communicating

Possible Hypotheses

The correct hypothesis is that you can predict horizontal range. Some students may incorrectly hypothesize that horizontal and vertical components of motion are dependent, and that the landing point is unpredictable.

Possible Materials

2 pieces of U-shaped channel (50 cm)
steel ball
masking tape
ring stand and clamp
meterstick
string (1 m)
washer
stopwatch or electronic timing apparatus with
 2 photogates and lights
paper cup

Possible Procedures

One procedure that works well is to use a U-shaped channel, a ring stand and clamp, and masking tape to construct horizontal and inclined tracks on a tabletop. A piece of string suspends a washer from the edge of the table to the floor, marking $x = 0$. The horizontal track should end at the edge of the table. Measure the distance (about 0.2 m) between two points along the horizontal track and use an electronic timer with photogates to measure the time it takes a ball released from a constant position on the inclined track to traverse the measured distance on the horizontal track. Use these data and the vertical distance to the floor from the end of the horizontal track to calculate the range. To determine the accuracy of the prediction, place a cup at the predicted distance and let the ball roll.

Teaching Suggestions

- Channels for the ball could be a flat chan-neled curtain rod, grooved rulers, or shelv-ing brackets. Cups for catching the ball should be fairly shallow.

- Have students work in pairs or in groups of three or four. Each student should collect some data. Be sure that students understand the operation of the timing apparatus and photogates. Using a stopwatch may result in less accurate results.

- Make sure that the lab tables are level. If necessary, students should use a level to adjust the track so that it is truly horizontal. An initial vertical component, even if small, can produce considerable deviations from the predicted results.

- Review calculation techniques, and make sure that the concepts involved are clear.

Data and Observations

[Sample data are given.]

Vertical distance from end of track to floor: 0.96 m

Distance interval chosen along the track: 0.20 m

Table 1

Trial	Time on ramp (s)	Horizontal velocity, v_x (m/s)	Time in air, t (s)	Predicted range (m)
1	0.152			
2	0.153			
3	0.154			
Calcu-lated	0.153	1.3	0.044	0.57

Analyze and Conclude

1. Answers will vary. The ball should land in the cup.

2. $t = \sqrt{\dfrac{2y}{g}}$

   $= \dfrac{2(6.0 \text{ m})}{9.80 \text{ m/s}^2} = 1.1 \text{ s}$

   $x = v_x t = (3.75 \text{ m/s})(1.1 \text{ s}) = 4.1 \text{ m}$

3. Answers will vary. Sample answer: The results would probably be different. The displacement of the sponge ball or Ping-Pong ball could be affected by the signifi-cant frictional resistance of the air.

Apply

1. $v_i = \dfrac{160\ 000 \text{ m/h}}{3600 \text{ s/h}} = 44.4 \text{ m/s}$

   $x = \dfrac{v_i^2}{g} (\sin 2\theta_i)$

   $= \dfrac{(44.4 \text{ m/s})^2}{9.80 \text{ m/s}^2} (2(0.891)(0.891)(0.454))$

   $= 145 \text{ m}$

7-3

Torques

Introduction

For a rotating object, the torque (τ) is the prod-uct of the vector quantity force (F) and lever arm (r). The lever arm is defined as the perpendicular distance between the axis of rotation and a line drawn in the direction of the force. If the sums of clockwise and counterclockwise torques about the axis are equal, the object will be in rotational equilibrium and will not turn about the axis. The resultant of a number of forces can be involved. In this investigation, two upward parallel forces, measured by spring balances, are used to balance a downward gravitational force. The upward forces will tend to cause rotation, producing a clockwise and a counterclockwise torque. By measurement of forces and lever arms, it is possi-ble to compare the torques involved.

Process Skills

observing and inferring, using numbers, inter-preting data, predicting, measuring in SI, analyzing

Preparation Hints

String or masking tape may be used in place of the meterstick clamps. The spring scales must be able to move freely so that students can make fairly accurate readings.

Materials

meterstick
2 spring scales, 5-N capacity or greater
3 meterstick clamps
500-g hooked mass
masking tape

Procedure Hints

Explain that, because torque is a vector quantity, by definition it includes direction in its specification. By mathematical convention, torques in a clockwise direction are negative and torques in a counterclockwise direction are positive.

Data and Observations

Table 1 (Sample data are given.)

	Balance a			Balance c		
Trial	Original reading (N)	Final reading (N)	Real reading (N)	Original reading (N)	Final reading (N)	Real reading (N)
1	0.85	3.15	2.30	0.75	3.10	2.35
2	1.00	3.95	2.95	0.60	2.55	1.95
3	1.30	4.90	3.60	0.35	1.55	1.20

Table 2 (Sample data are given.)

Trial	Distance AB (m)	Distance BC (m)	Clockwise torque (N·m)	Counterclockwise torque (N·m)
1	0.45	0.45	−1.04	1.06
2	0.30	0.45	−0.89	0.88
3	0.15	0.45	−0.54	0.54

Analysis and Conclusions

1. Because the system in each trial was not rotating about the axis, the sum of the clockwise torques and the sum of the counterclockwise torques was zero.

2. As the length of the lever arm increased, the force exerted decreased.

3. Trial 1: 2% difference; Trial 2:1% difference; Trial 3: 0% difference

4. The counterclockwise torque should have a value equal to and opposite that of the clockwise torque.

5. In equilibrium, there is no rotational movement; the sum of all torques is, therefore, zero. Because $W = \tau \Delta\theta$, if $\Sigma\tau = 0$, then $W = 0$. No work was done.

Extension and Application

1. The total weight supported by the piers of the bridge is the weight of the bridge plus the weight of the person, or 980 N + 540 N = 1520 N. Because the bridge is not rotating, it is in rotational equilibrium. Therefore, the sum of the clockwise torques equals the sum of the counterclockwise torques. If the force F_1 at one end is the pivot point for possible rotation,

   $F_1 d = F_{bridge}d + F_{person}d$, where d is the distance from the pivot point. Solve for F_1.

   $(F_1)(5.0 \text{ m}) = (980 \text{ N})(2.5 \text{ m}) + (540 \text{ N})(2 \text{ m})$

   $F_1 = (2450 \text{ N·m} + 1080 \text{ N·m})/5 \text{ m}$

   Thus, $F_1 = 710$ N at the end farther from the person, and $F_2 = 1520$ N − 710 N = 810 N at the end closer to the person.

8-1

Kepler's Laws

Introduction

Planetary motions and positions can be described and related quantitatively. The planets sweep out equal areas in equal times and have periods that are related to their average distance from the sun. They travel in elliptical orbits with the sun at one focus. The eccentricity, e, of an elliptical orbit is a measure of how much it departs from a circular shape. It is equal to the difference between the lengths of the longest and the shortest radius, divided by the length of the major axis. The eccentricity can theoretically range from zero (for a perfectly circular orbit) to one (for a maximally long and flattened ellipse). For Earth, e equals 0.0167, whereas for Mercury, with its more eccentric orbit, e equals 0.206.

Polar graph paper can be used to plot the values of the radius vectors for the positions of a planet, using tables of heliocentric data. The orbit can then be drawn, and the various quantitative relationships explored.

Process Skills

formulating models, comparing and contrasting, interpreting data, defining operationally, using numbers

Materials

polar graph paper
sharp pencil
metric ruler

Procedure Hints

Students should plot data precisely. When plotting radius vectors, they should strive for an error of no more than the width of a sharpened pencil lead. The better the plot, the easier it will be to interpret the result.

Data and Observations

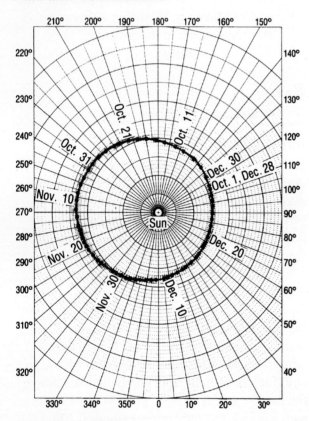

The orbit of Mercury

Analysis and Conclusions

1. Yes. Mercury's orbit is an ellipse, with the center of the sun at one focus.

2. The area for the section of Mercury's orbit between December 20 and December 30 is
   area = $(61°/360°)(\pi)(0.318 \text{ AU})^2$
   $= 0.054 \text{ AU}^2/10 \text{ days}$

3. Answers will vary. Two choices:
   Oct. 11–21,
   area = $(39°/360°)(\pi)(0.396 \text{ AU})^2$
   $= 0.053 \text{ AU}^2/10 \text{ days}$
   Nov. 10–20,
   area = $(29°/360°)(\pi)(0.457 \text{ AU})^2$
   $= 0.053 \text{ AU}^2/10 \text{ days}$

4. The average area should be about 0.053 $\text{AU}^2/10$ days. Students should have errors of less than 5% if their graphs are carefully plotted. Yes, the orbit supports Kepler's law of areas.

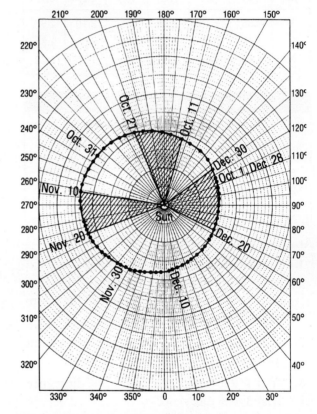

Three areas of the orbit of Mercury

5. Average radius = $(0.307 \text{ AU} + 0.467 \text{ AU})/2$
   $= 0.387 \text{ AU}$

$$\frac{T_M^2}{T_E^2} = \frac{r_M^3}{r_E^3}$$

$$T_M^2 = \frac{(0.387 \text{ AU})^3(365.25 \text{ days})^2}{(1.0 \text{ AU})^3}$$

$$T_M = 88 \text{ days}$$

6. From the graph, the period = 30 days + 30 days + 28 days = 88 days. The value from question 5 is the same (88 days). Thus, the results from the graph are consistent with Kepler's law of periods.

Extension and Application

1.
$$T_x^2 = \frac{(1.0 \text{ AU})^3 (1.0 \text{ yr})^2}{(1.0 \text{ yr})^2} = (1.0 \text{ yr})^2$$

$$T_x = 1.0 \text{ yr}$$

2. The figures below show the orbit of Mars and three areas of the orbit of Mars. The area per day is 0.01 AU². The orbit is an ellipse. Yes, Kepler's law of areas applies.

The orbit of Mars

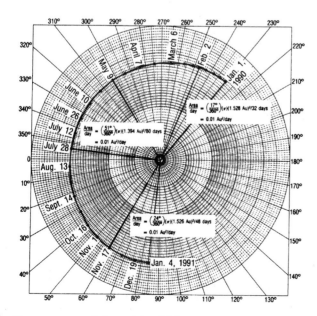

Three areas of the orbit of Mars

9-1

Conservation of Momentum

Introduction

The momentum, p, of an object is a vector quantity that is equal to the product of the mass and the velocity. According to the law of conservation of momentum, the momentum of a closed, isolated system remains constant. The law applies not simply to individual objects, but also to objects that collide with one another or an exploding projectile such as a fireworks aerial display. In this investigation, the system was initially at rest and then both carts moved in different directions after the spring mechanism was released. Momentum before and after were equal.

Process Skills

observing, interpreting data, defining operationally, using numbers, measuring in SI

Preparation Hints

The motion detector will not accurately measure the cart when it is closer than 0.4 meters. Remind students to keep the motion detector and cart at least 1.5 meters away initially.

Materials

safety goggles
2 collision carts, one with a spring-release mechanism
balance
meterstick
CBL unit
ultrasonic motion detector
link cable
graphing calculator with PHYSICS CBL software program
set of masses
masking tape
two 5 × 7 index cards
ring stand
clamp

Procedure Hints

- The masses need to be secured to the cart so they do not move during the spring-mechanism release. Sometimes it is helpful to place the masses at the end of the cart against the edge facing the adjacent cart. They are less apt to move in this location.

Data and Observations

Table 1 (Sample data are given.)

	Mass (kg)	Mass Added (kg)
Cart 1	791	—
Cart 2	840	750

Table 2 (Sample data are given.)

	Begin t (s)	End t (s)	Δt (s)	Begin d (m)	End d (m)	Δd (m)	Average velocity (m/s)
Cart 1-trial 1	1.11	1.98	0.87	1.41	1.02	0.39	0.44
Cart 1-trial 2							
Cart 2-trial 1	1.23	1.83	0.60	1.36	1.24	0.12	0.2
Cart 2-trial 2							

Analysis and Conclusions

1. Velocity values in Table 2. Answers will vary.

2. Answers will vary. However, momentum of the carts will be approximately equal.

3. Zero (The velocity of each cart was zero.)

4. No. After the mechanism was released, the carts both moved off in opposite directions.

5. The final momentum should be approximately zero. The momentum of one of the carts should be negative, since it was moving in the opposite direction.

Extension and Application

1. If you let your hands move backwards as you catch the ball, the transfer of momentum will occur over a longer period of time. This extension will relieve the stinging.

2. The final momentum of the incident ball and the two target balls will equal the initial momentum of the incident ball. The total momentum of a closed, isolated system always remains constant no matter how many objects are in the system. This relationship is shown by:

$$p_{A1} = p_{A2} + p_{B2} + p_{C2}.$$

3. Mass cancels out, and the equation is a form of the Pythagorean theorem, $v_{A1}^2 = v_{A2}^2 + v_{B2}^2$, for a right triangle. Therefore, if this equality holds, the angle between the velocity vectors after the collision must be a right angle.

9-2

Angular Momentum

Introduction

For an object that is rotating, the relationship between torque (τ), rotational inertia (I), and angular acceleration (α) is as follows:

$$\tau = I\alpha$$

The torque can be considered a rotating force applied to the object, and the rotational inertia is the resistance to changes in the object's angular velocity (ω). The angular velocity is in turn related to the rotational inertia and to the angular momentum (L) by the following equation:

$$L = I\omega$$

It is possible to carry out a simple experiment to study the concept of angular momentum and its conservation. A rotating stool or platform can be used to observe the effects of arm position on rotational inertia and thus on rate of spin. A spinning gyroscopic bicycle wheel can also be used to observe what happens when attempts are made to tilt the wheel. In these ways, the conservation of angular momentum can be verified.

Process Skills

observing and inferring, communicating, analyzing, defining operationally, experimenting

Preparation Hints

A gyroscope can be substituted for the spinning wheel. However, the results will not be as dramatic for students as when they try to change the angular momentum of the rotating bicycle wheel.

Materials

gyroscopic bicycle wheel
rotating stool or rotating platform
2 masses, 3 kg each
safety goggles

Procedure Hints

- This laboratory activity reinforces the supplemental lesson on rotary motion and angular momentum in the booklet *Advanced Concepts* in the TRP.

- If you do not use the TRP lessons on rotational inertia and angular momentum, this experiment can be an enrichment activity after students complete the lessons on conservation of momentum for translational motion.

Data and Observations

1. Observations of student holding masses on spinning stool or platform:

   When the student pulls in his or her arms, the rate of rotation increases. When the arms are extended again, the rate of rotation decreases.

2. Observations of effects of tilting the shaft of a stationary wheel upward:

   Other than being somewhat heavy, the wheel easily tilts upward.

3. Observations of effects of tilting the shaft of a rotating wheel upward:

   The rotating wheel resists efforts to tilt it upward. When students attempt to tilt it upward quickly, the wheel suddenly swerves to the right.

4. Observations of effects of rotating the spinning wheel:

   When the wheel rotates to the right, the stool rotates to the left.

Analysis and Conclusions

1. Since the angular momentum must be conserved, the product of $I\omega$ must be constant. When the student's arms are pulled in, the rotational inertia decreases. If I has a smaller value, ω must increase to conserve the angular momentum of the system: $I_1\omega_1 = I_2\omega_2$

2. Answers may vary. One condition could be an external torque applied to the system. For example, another student might give the stool an extra spin, thus increasing the rate of rotation and the angular momentum of the student-stool-masses system.

3. When the shaft tilts, its angular momentum is changed by a small amount, ΔL. In order to cause a small change in L during an interval of time, $\Delta L/\Delta t$, a torque must be applied. If the torque is not applied in the same direction as ΔL, the wheel will swerve off to one side or another (depending on the direction of rotation).

4. For angular momentum to be conserved, the stool must rotate to the left when the wheel rotates to the right. (Students may report the opposite results if their wheels were spinning in the opposite direction.)

Extension and Application

1. Many students will say that the two objects should reach the bottom of the incline at the same time. However, the solid cylinder will reach the bottom first. The thin rim wheel has greater rotational inertia than does the cylinder, and it will roll down the incline at a slower rate.

2. The moment the diver leaves the springboard, he is a closed, isolated system with a given angular momentum that cannot change while he is in the air. When he goes into the tuck position, his rate of rotation increases and his rotational inertia decreases. When he goes into the layout position prior to hitting the water, he increases his rotational inertia and decreases his rate of rotation. Angular momentum is conserved throughout the dive.

3. Acceleration due to gravity exerts an external torque on the child-swing system. Without this torque, swinging would be impossible.

10-1

Pulleys

Introduction

In an ideal machine, one lacking friction, all the energy is transferred, and the work input of the system equals the work output. The work input equals the force times the distance that the force moves, $F_e d_e$. The work output equals the output force (load) times the distance it is moved, $F_r d_r$. The ideal mechanical advantage, IMA, of the pulley system can be found by dividing the distance the force moves by the distance the load moves. Thus $IMA = d_e/d_r$. The ideal machine has a 100% efficiency. In the real world, however, the measured efficiencies are less than 100%. Efficiency is found using the following equation:

$$\text{Efficiency} = \frac{\text{Work output}}{\text{Work input}} \times 100\%.$$

Process Skills

observing and inferring, using numbers, interpreting data, measuring in SI, defining operationally, experimenting

Materials

2 single-mounted pulleys
2 double-mounted pulleys
set of hooked metric masses
spring scale, 5-N capacity or greater
pulley support
heavy string (2 m)
meterstick

Preparation Hints

- You may need to relate to students the information in the Introduction above.

Procedure Hints

- Grade the results reasonably. There are several sources of error in doing the laboratory activity, mostly in getting accurate spring scale readings. Consequently, results may not show very close agreement. Demonstrate the proper use of the spring scale in each arrangement. Emphasize the necessity for using the least pull required to keep the mass moving upward at a slow and steady rate. Tell students to consider other potential sources of error and to take precautions that will minimize their effect on results.

- Demonstrate the measurement of F_g, h, F, and d. Students often have difficulty measuring d, the distance the force moves.

- If time and equipment permit, you may want to let students experiment with the pulley systems they have designed in response to the Extension and Application suggestion.

Data and Observations

Table 1 (Sample data are given.)

Pulley arrange-ment	Mass raised (kg)	Weight (F_g) of mass (N)	Height (h) of mass (N)	Force (F) of spring scale (N)	Distance (d) through which force acts (m)
1	0.50	4.9	0.25	5.0	0.27
2	0.50	4.9	0.25	2.8	0.50
3	1.0	9.8	0.25	5.0	0.53
4	1.0	9.8	0.25	2.7	1.02

Table 2 (Sample data are given.)

Pulley arrange-ment	Work output ($F_g h$) (J)	Work input (Fd) (J)	IMA (d_e/d_r)	Number of lifting strands	Efficiency %
1	1.2	1.4	1	1	86
2	1.2	1.4	2	2	86
3	2.5	2.7	2	2	93
4	2.5	2.8	4	4	89

Analysis and Conclusions

1. Inaccuracies in reading the spring scale, accuracy of the spring scale, and friction are the largest causes of error.

2. The force gets smaller as the mechanical advantage gets larger.

3. The IMA and the efficiency should not be affected.

4. The IMA increases from 1 to 2 pulleys or from 3 to 4 pulleys, but not when going from 2 to 3 pulleys. The efficiency may or may not decrease.

5. The values for IMA and the number of supporting strands should be the same.

6. The amount of work is not reduced, but the force that must be applied is reduced. Considering the losses of energy due to friction, more, not less, work usually must be done.

Extension and Application

1.

To boat

11-1

Conservation of Energy

Introduction

The law of conservation of energy states that energy can be neither created nor destroyed but simply changes form, the total quantity remaining constant. The law can be applied to situations involving motion along an inclined plane. The forces involved when a block is pulled up or allowed to slide down the plane can be determined, and comparisons of the work done and the potential energy can then be carried out. The total work in pulling in such a case must be done against both gravity and friction, and is equal to the increase in the gravitational potential energy, mgh. The changing of the path, or (in the case of this investigation) the angle of the tilt, should have no effect on the work done against gravity, thus confirming the law of conservation of energy.

Process Skills

observing and inferring, measuring in SI, using numbers, interpreting data, analyzing, defining operationally

Preparation Hints

A nonstick cooking spray can be substituted for the silicone spray.

Materials

inclined plane
smooth wood block with hook in one end
spring scale
meterstick
heavy string (1 m)
silicone spray

Procedure Hints

- Emphasize the importance of moving the block up and down the incline at constant speed. If the block accelerates, the scale reading will be wrong.

- The quantity $F_\parallel d$ is the work done going up the frictionless inclined plane. $F_{up}d$ represents the work input, including the work done against friction.

Data and Observations

Weight of the block: 6.0 N

Table 1 (Sample data are given.)				
Trial	Length, d (m)	Height, h (m)	F_{up} (N)	F_{down} (N)
1	0.96	0.46	5.0	1.0
2	0.73	0.46	5.5	2.0
3				

Table 2 (Sample data are given.)				
Trial	$F_\parallel$ (N)	Work done without friction, $F_\parallel d$ (J)	Potential energy, mgh (J)	Work input, $F_{up}d$ (J)
1	3.0	2.9	2.8	4.8
2	3.8	2.8	2.8	4.0
3				

Analysis and Conclusions

1.
$$F_{up} = F_\parallel + F_f$$
$$F_{down} = F_\parallel - F_f$$
$$F_{up} + F_{down} = 2F_\parallel$$
$$F_\parallel = (F_{up} + F_{down})/2$$

2. Trial 1: $F_\parallel$ = (5.0 N + 1.0 N)/2 = 3.0 N
Trial 2: $F_\parallel$ = (5.5 N + 2.0 N)/2 = 3.8 N

3. Trial 1: W = (3.0 N)(0.96 m) = 2.9 J
Trial 2: W = (3.8 N)(0.73 m) = 2.8 J

4. U = (6.0 N)(0.46 m) = 2.8 J

5. In the absence of friction, the energy within the system is conserved. Students should have nearly the same values for work and potential energy.

Extension and Application

1. Answers will vary.

Trial 1

$\sin \theta = 0.46/0.96$

$\theta = 29°$

$F_\parallel = W \sin \theta = (6.0 \text{ N})(0.48) = 2.9 \text{ N}$

Trial 2

$\sin \theta = 0.46/0.73$

$\theta = 39°$

$F_\parallel = W \sin \theta = (6.0 \text{ N})(0.63) = 3.7 \text{ N}$

Most students will suggest that they may have pulled too hard, causing some acceleration of the block, or they didn't pull parallel to the plane, thus increasing the force reading.

2. Since the hill is frictionless, the gravitational force converts the stored potential energy of the skier at the top of the hill to kinetic energy.

$$mgh = \tfrac{1}{2}mv^2$$

$$(60.0 \text{ kg})(9.80 \text{ m/s}^2)(10.0 \text{ m}) = \tfrac{1}{2}(60.0 \text{ kg})v^2$$

$$v^2 = 196 \text{ m}^2/\text{s}^2$$

$$= 14 \text{ m/s}$$

12-1

Specific Heat

Introduction

The specific heat of a material is the amount of energy it must absorb per kilogram to increase in temperature by 1 kelvin. The relationship between specific heat (c), heat gained or lost (Q), mass (m), and temperature change (ΔT) is as follows:

$$c = \frac{Q}{m\Delta T}.$$

The law of conservation of energy may be applied to determine the specific heats of various metals. The temperature changes produced by the immersion of heated metals in a simple calorimeter containing water can be measured. The results of such an investigation provide the data needed to solve the equation that will yield the specific heats, allowing direct comparison of the characteristic values for the various metals tested.

Process Skills

observing and inferring, interpreting data, measuring in SI, using numbers, comparing and contrasting

Preparation Hints

- Mercury thermometers are dangerous and should be removed from school laboratory supply stock. Use only alcohol thermometers to avoid the risk of a mercury spill.

- The old-style calorimetry cups (typically aluminum) can be used. If this type is used, then the heat lost by the hot metal equals the heat gained by the water plus the heat gained by the calorimeter.

- If lead is used, the mass must be large enough (300–400 g) to cause a measurable temperature change.

Materials

CBL unit
link cable
glass rod
graphing calculator with PHYSICS CBL software program
temperature probe
string (60 cm)
safety goggles
250-mL beaker
polystyrene cup
hot plate (or burner with ring stand, ring, and wire screen)
tap water
balance
specific heat set (brass, aluminum, iron, lead, copper, etc.)

Procedure Hints

- Review with students the values for specific heat of common substances found on page xv. They may note that the value for water is very high. Explain that this property of water makes it a particularly good coolant for car engines.

Data and Observations

Table 1 (Sample data are given.)

	Trial 1	Trial 2
Type of metal	brass	aluminum
Mass of calorimeter cup (kg)	0.0016	0.0016
Mass of calorimeter cup and water (kg)	0.1203	0.1251
Mass of metal (kg)	0.2741	0.0905
Initial temperature of room-temperature water (°C)	22	23
Temperature of hot metal (°C)	100	100
Final temperature of metal and water (°C)	35	33.5

Table 2 (Sample data are given.)

	Trial 1	Trial 2
Mass of room-temperature water (kg)	0.1187	0.1235
ΔT metal (°C)	65	66.5
ΔT room-temperature water (°C)	13	10.5

Analysis and Conclusions

1. See Table 2.

2. Trial 1
$$Q_{\text{gained by water}} = (0.1187 \text{ kg})(13°\text{C})(4180 \text{ J/kg·°C})$$
$$= 6450 \text{ J}$$

   Trial 2
$$Q_{\text{gained by water}} = (0.1235 \text{ kg})(10.5°\text{C})(4180 \text{ J/kg·°C})$$
$$= 5420 \text{ J}$$

3. Trial 1
$$C_{\text{brass}} = 6450 \text{ J}/(0.2741 \text{ kg})(65°\text{C})$$
$$= 362 \text{ J/kg·°C} = 362 \text{ J/kg·K}$$

   Trial 2
$$C_{\text{aluminum}} = 5420 \text{ J}/(0.0905 \text{ kg})(66.5°\text{C})$$
$$= 901 \text{ J/kg·°C} = 901 \text{ J/kg·K}$$

4. Trial 1
$$\text{relative error} = [(376 - 362) \times 100\%]/376$$
$$= 4\%$$

Trial 2
$$\text{relative error} = [(903 - 901) \times 100\%]/903$$
$$= 0.2\%$$

Student errors should be less than 15%. If samples are lead or have very small mass, errors will be larger because the temperature changes in these situations will not be great enough to measure.

5. The thermometer may not be reading accurately. (However, the difference in temperature should be accurate.) Heat may have radiated to the environment from the top of the cup. Some hot water from the sample may have dripped into the calorimeter.

Extension and Application

1. Students should be able to identify the substance by its specific heat. Be sure that the unknown has a large enough mass or the temperature change will be insignificant, and errors will be large.

2. $$C_{\text{substance}} = \frac{m_{\text{water}} \Delta T_{\text{water}} (4180 \text{ J/kg·K})}{m_{\text{substance}} \Delta T_{\text{substance}}}$$

$$= \frac{(150.0 \text{ g})(7.1°\text{C})(4180 \text{ J/kg·K})}{(100.0 \text{ g})(67.9°\text{C})}$$

$$= 656 \text{ J/kg·K}$$

From the table on page xv, this value for specific heat is closest to that of glass.

12-2

Efficiency of Solar Collectors

Introduction

Students will plan and execute a procedure for comparing the heat-collecting abilities of four materials. They will build a simple device for collecting solar energy and can use a computerized system to gather and analyze data. A graph of the data will help them compare the four materials. In carrying out this activity, they will be exploring topics relating to temperature and thermal energy.

Process Skills

hypothesizing; measuring; using variables, constants, and controls; comparing and contrasting; collecting and organizing data; making and using graphs

Possible Hypotheses

Students may legitimately hypothesize any of the materials as most efficient.

Possible Materials

large coffee can with hole punched in side and clear plastic lid

small coffee can with hole punched in side and clear plastic lid with center cut out

sheet of plastic wrap

sheet of wax paper

piece of polyethylene

small pane of glass

shredded newspapers

CBL unit

TI-82 CBL graphics calculator with unit-to-unit link cable

TI temperature probe

TI-Graph Link

colored pencils

Possible Procedures

The following procedure works well. Stretch the first material to be tested, such as the piece of plastic wrap, across the top of the small can, and put the cutout lid on top. Place the small can in the large one, packing shredded newspaper between the two cans and lining up their holes.

Large coffee can

Small coffee can with cut-out lid

Temperature probe

Shredded newspapers

Insert the temperature probe though the holes. The probe will measure the temperature of the air inside the small can. Turn on the CBL and start the program TEMP on the TI-82. At the prompt, enter the channel number to which the probe is connected and the number of samples

desired in 10 min. Sample every 4 s. Put the cans in the sun and immediately press ENTER on the TI-82 to start collecting data. Record a sampling of times and temperatures in a table.

A STAT PLOT of time (seconds) versus temperature (°C) for the setup will be displayed. Find the initial and final temperature values using TRACE on the STAT PLOT. Save the temperature data from L4 and the time data from L5 to a computer, using the TI-Graph Link.

Repeat the procedure, using the other materials as covers.

Students may construct the heat-collecting device differently. Some students may prefer to collect their data without relying on computerized devices.

Teaching Suggestions

- This lab requires significant preparation of equipment (include paint-drying time) and may require training students how to use CBL and the TI-82. The lab should be conducted on a sunny day.

- To prepare the equipment: Use a hammer and nail to punch a hole in the side of each can for the temperature probes. Smooth down any jagged edges. Paint the inside of each small can flat black and allow it to dry. Cut the center out of the plastic lid of the small can, leaving 1 cm all around. Tape the edges of the glass panes.

- Be sure that students know how to use the various devices. The CBL unit connects to the TI-82 calculator with a unit-to-unit link cable, using the I/O ports on the bottom edge of each unit. The cable should be pressed in firmly. The temperature probe should be connected to channel 1 (CH1) on the top edge of the CBL. The TEMP.82P program, listed in the appendix of the *CBL System Guidebook*, should be downloaded or entered.

- Groups of four students work well, with each student carrying out most of the procedure for one material. Two students can set up the TI CBL equipment while the other two construct the solar collector.

- Tell students that, for each sample, the CBL displays a STAT PLOT of time (seconds) versus temperature (°C). Pressing TRACE on the STAT PLOT gives the initial and final temperatures. The temperature data from L4 and the time data from L5 should be saved to a computer using the TI-Graph Link. Data for hand graphing purposes should be collected about 10 times for each material, preferably 20–60 s after the initial readings.

- Caution students not to attempt to put the glass in the lid of the outer can. Also tell them to keep the end of the temperature probe away from the metal of the can.

- Review students' plans, and accept reasonable procedures.

Data and Observations

Actual times and temperatures will vary; however, the pattern of increase for the four materials should be similar.

Table 1 (Sample data are given.)				
Temperature (°C)				
Time (s)	Plastic wrap	Wax paper	Polyethylene	Glass
0	20	20	20	20
20	28	24	26	29
28	30	25	28	31
32	32	26	30	33
36	33	28	31	34
40	34	29	32	36
44	35	30	33	38
48	35	31	33	38
60	35	31	34	39

Analyze and Conclude

1. The graphs should resemble the one below. Students' plots should correspond fairly closely to STAT PLOT, with similar slopes.

2. Glass is most efficient. Wax paper is least efficient. The curve for glass showed the greatest increases in temperature, indicating the most heat. That for wax paper showed the least increases.

3. Answers will vary. Students will probably choose to build a glass shelter because it collects heat best. They may select other materials for practical reasons, such as ease of construction.

Apply

1. Polyethylene would be cheaper to build with. However, glass would likely be cheaper in terms of energy expenses, given glass's greater ability to collect heat. Efficiency could be improved by methods such as doubling layers of material.

13-1

Archimedes' Principle

Introduction

Archimedes' principle states that the buoyant force on an object is the difference between the object's weight in air and its weight in water. This can be explained by Newtonian mechanics. The downward weight of the object equals the object's mass times the gravitational acceleration. This is measured by the spring scale in air. An object weighs less in water than in air because the higher pressure in the water column below the object results in a net upward force on the object. In essence, the sum of the object's submerged weight and the buoyant force equals the object's weight in air.

Misconception: It may be necessary to reiterate that the buoyant force depends on the weight of the displaced fluid, not the weight of the submerged object.

Process Skills

observing and inferring, comparing and contrasting, measuring in SI, using numbers, interpreting data, communicating

Preparation Hints

- Be sure that the spring scales work properly. If they stick, a large error is introduced into the results.

- Each laboratory group could be given different masses, such as 200 g, 400 g, and 1000 g. To prevent damage to the spring scale, be sure it can measure the larger masses.

Materials

500-mL beaker
spring scale, 5-N capacity or greater
500-g hooked mass
100-g hooked mass
polystyrene cup
paper towel

Procedure Hints

- Review the equation for buoyant force, $F_b = \rho V g$. This will help students understand that the buoyant force is related to the weight of the volume of fluid displaced, not the weight of the submerged object.

- Because the volume markings on beakers are highly variable and often inaccurate, students will obtain better results if they first calibrate their beaker. They can do this by pouring measured volumes of water from a graduated cylinder into the beaker and marking the levels on a piece of tape affixed to the side of the beaker.

Data and Observations

Table 1 (Sample data are given.)	
Weight of 500-g mass in air	4.9 N
Weight of 500-g mass immersed in water	4.2 N
Volume of water in beaker	300 mL
Volume of water in beaker with 500-g mass immersed	375 mL

Table 2 (Sample data are given.)	
Volume of water in beaker	300 mL
Volume of water with 100-g mass immersed	320 mL
Volume of water with 100-g mass in polystyrene cup	400 mL

Analysis and Conclusions

1. $F_{buoyant} = 4.9 \text{ N} - 4.2 \text{ N} = 0.7 \text{ N}$

2. Volume displaced $= 375 \text{ mL} - 300 \text{ mL}$
   $= 75 \text{ mL};$
   75 mL of water has a mass of 75 g.
   $$F = (0.075 \text{ kg})(9.8 \text{ m/s}^2) = 0.74 \text{ N}.$$
   Explanations will vary, but may include errors in measuring volume and force. Students might also note the effect of air bubbles that adhere to the surface of the mass.

3. It rose. Combined, the 100-g mass and boat have an average density less than that of water, and the boat (with mass) floats. The volume of water displaced by the submerged mass alone was 20 mL; the 100-g mass in the boat displaces 100 g of water. Because 100 g of water has a volume of 100 mL, this is the difference between the observed volumes.

Extension and Application

1. When Tim and Sally fall into the water, they displace a volume of water that weighs less than they do; the volume of water displaced is equal to the buoyant force acting on them. Because they displace less water while submerged than when floating, the pool water level drops when they fall off the raft.

2. $\dfrac{0.92 \text{ g/cm}^3}{1.03 \text{ g/cm}^3} \times 100\% = 89\%$ below the surface, so 11% is above the surface.

14-1

Ripple Tank Waves

Introduction

A simple device called a ripple tank can be used to investigate various properties of waves. The ripple tank can be set up to permit observations of phenomena including reflection, refraction, diffraction, and constructive and destructive interference of water waves. Changes of medium can be simulated by changing the depth of the water. Both straight pulses and circular waves can be generated, and barriers can be used in various orientations to reveal the effects on wave behavior. The ripple tank also allows for observation of wave superposition. The overall results obtained through use of the device are applicable to the study of both longitudinal and transverse waves.

Process Skills

observing and inferring, formulating models, analyzing, interpreting data, predicting, defining operationally

Preparation Hints

- A straight-wave generator can be substituted for the dowel, but students often have trouble adjusting it to work effectively. The dowel will usually generate a straight wave.

- This is a long experiment and will probably require two lab periods.

- Several point-source generators can be assembled and shared among lab groups.

- If you use paraffin blocks, separate and store them in a cool place because they will melt in warm weather.

Materials

ripple tank
light source
dowel rod (30 cm long × 2 cm diameter)
3 wood or paraffin blocks
metal strip (parabola) or rubber hose
meterstick
protractor
large sheets plain white paper
eyedropper
glass plate
12 washers or pennies for glass supports
2-point-source wave generator
0–15 VDC variable power supply

Procedure Hints

If students have trouble producing good images, suggest that they either increase or decrease the water depth in the ripple tank.

Data and Observations

Table 1 (Sample data are given.)
1. Observations of straight pulses striking a straight barrier at 0° and at another angle:
Pulses strike the barrier and bounce off. If they strike straight on, 0° to the normal, they reflect at an angle of 0°. If they strike the barrier at an angle to the normal, they reflect at an angle equal to the normal.
2. Sketch of incoming pulses and reflected pulses:
3. Statement of the relationship between the angle of incidence and the angle of reflection:
The angle of incidence equals the angle of reflection.
4. Description of the shape of the reflected pulse from a concave surface:
Straight waves reflected from a concave surface converge at a common point.
5. Description of the reflection of circular waves originating at the focus:
The waves originating at the focal point of a concave barrier are reflected as straight waves.

Table 2 (Sample data are given.)

Observations of refraction of straight waves:

Glass plate

Straight-wave generator

1   2

Table 3 (Sample data are given.)

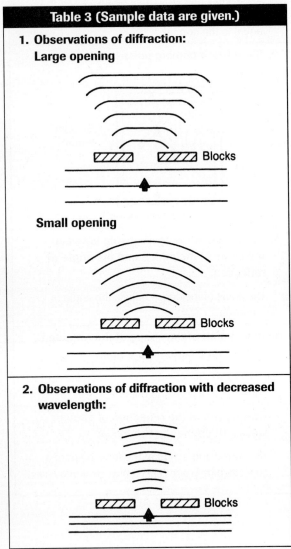

1. Observations of diffraction:
    Large opening

Blocks

Small opening

Blocks

2. Observations of diffraction with decreased wavelength:

Blocks

Table 4 (Sample data are given.)

1. Interference pattern:

Antinodal lines   Nodal lines

2. Interference pattern (higher frequency than item 1):

Analysis and Conclusions

1. The angle of incidence equals the angle of reflection.

2. As a wave passes from deep to shallow water, its velocity decreases, wavelength decreases, and frequency remains the same.

3. Diffraction becomes more noticeable when the slit opening is smaller than the wavelength.

4. As the wavelength decreases, the angle of spread of the diffracted waves becomes smaller.

5. As the wavelength increases, the emerging interference pattern and nodal lines radiate outward at a greater angle. The smaller the wavelength, the greater the number of nodal (and antinodal) lines.

6. The diffracted waves will have an interference pattern like that produced by a two-point source.

7. Light is diffracted through the opening and creates an interference pattern with nodal and antinodal lines.

Extension and Application

1. The ocean waves bend and slow down as they pass over a submerged reef or sandbar.

2. If you understand that the angle of reflection equals the angle of incidence, you can choose a good position to hit the returning ball (platform tennis or racquet ball) or line up your next shot (pool).

3. The waves spread out, forming an interference pattern similar to the one produced by the two-point source. This result should match students' predictions from question 6.

14-2

Velocity, Wavelength, and Frequency in Ripple Tanks

Introduction

This lab extends the observation of wave properties begun in Lab 14-1. The present lab focuses on the general relationship between three essential characteristics of waves: velocity of propagation, wavelength, and frequency. The medium can be kept constant and the frequency varied, with measurement of the resulting wavelengths. A ripple tank and stoboscope can be used in determining the various values. This investigation permits observation of the following general relationship between wave velocity (v), frequency (f), and wavelength (λ):

$$v = f\lambda$$

It also makes possible the verification of the constancy of wave velocity in a given medium (water of constant depth).

Process Skills

observing and inferring, using numbers, measuring in SI, interpreting data, hypothesizing, analyzing data, experimenting

Preparation Hints

- If students have never used stroboscopes before, explain how they work.

- Straight-wave generators may be substituted for the point source. Students may have difficulty using one side of the wave generator; if the plunger is off center or leaning, waves of poor quality are produced.

Materials

ripple tank
light source
2-point-source wave generator
0–15 VDC variable power supply
stopwatch or watch with second hand
meterstick
stroboscope
large sheets plain white paper
masking tape
2 rulers

Procedure Hints

- Encourage students to be patient. The activity will work well if they carefully follow each step.

- If the images are not clear and sharp, have students add or remove a little water. The wave generator must be able to move up and down without touching the bottom of the ripple tank.

Data and Observations

	Table 1 (Sample data are given.)		
Trial	Width of five wavelengths (m)	Time for ten revolutions (s)	Time for one revolution (s)
1	0.105	4.1	0.41
2	0.116	4.8	0.48
3			

Table 2 (Sample data are given.)		
Length of object (cm)	Length of shadow (cm)	Magnification ratio (length of shadow/ length of object)
12	24	24/12 = 2

Table 3 (Sample data are given.)				
Trial	Observed wavelength (m)	Corrected wavelength (m)	Frequency (Hz)	Velocity (m/s)
1	0.021	0.011	9.72	0.11
2	0.023	0.012	8.33	0.10
3				

Analysis and Conclusions

1. The frequency of the wave was manipulated, and the wavelength responded to the change in frequency.

2. The velocities should show fairly good agreement. They should vary no more than 10–15%. The velocity should be the same but, with the procedure used, may show some variation.

3. Yes. Since the velocities calculated from $f\lambda$ are nearly equal, velocity of a wave in a given medium is constant.

Extension and Application

1. Frequency is $1/T$, with T the period of the wave, and the wavelength is the distance between waves, so $v = \lambda/T$, which is equivalent to $v = \Delta d/\Delta t$.

2. $f = v/\lambda$
   $= (217 \text{ m/s})/(3.5 \text{ m}) = 62 \text{ Hz}$
   $T = 1/f$
   $= 1/(62 \text{ Hz}) = 0.016 \text{ s}$

15-1

The Sound Level of a Portable Radio or Tape Player

Introduction

The actual intensity (I) of sound does not correspond to the perception of loudness by the human ear. A tenfold increase in intensity is perceived as only a doubling of loudness. Instead of actual intensity, a quantity called relative inten-sity, which corresponds more closely to perception, is often measured. A unit called the decibel is often used as a measure of this relative intensity. Each 10-decibel increase is perceived as a doubling of loudness.

The decibel levels of the sounds produced by an electronic source can be measured in various ways. In this investigation, a sound-level meter will be used in conjunction with a radio or tape player. The sound source can be adjusted to various volume settings, at each of which the decibel level can be measured. That will also make possible an examination of the relationship between sound intensity and perceived loudness, as well as an exploration of the issue of exposure time and hearing safety.

Process Skills

using numbers, communicating, interpreting data, measuring in SI, predicting, defining operationally, recognizing cause and effect

Preparation Hints

- To improve results, record a 1000-Hz tone on a cassette tape to serve as the sound source.

- Make arrangements with your school nurse to have students' hearing tested. Have students copy their hearing records. They should look for changes in hearing sensitivity and for trends, such as gradual hearing loss in one or both ears.

Materials

portable radio or tape player, with earphones
cassette tape of music
masking tape
sound-level meter
white correction fluid
graph paper

Procedure Hints

Have students investigate various areas in your school, such as the office, copy center, maintenance area, and cafeteria, recording sound levels and types of equipment producing the sounds. Have them compare the measured levels with the time exposure permitted at these levels.

Data and Observations

Table 3 (Sample data are given.)			
Sound level (dB)			
Setting	Trial 1	Trial 2	Trial 3
0	—	—	
1	60	67	
2	60	79	
3	66	90	
4	70	96	
5	74	101	
6	77	106	
7	83	109	
8	86	111	
9	90		
10	94		

Table 4 (Sample data are given.)	
Volume setting for listening	8
Estimated time at preferred volume setting	1 h

Analysis and Conclusions

1. Answers will vary depending on the type of volume control in each radio or tape player. Some will be linear and some will be logarithmic.

This graph was produced with the Lab Partner graphing program.

2. Answers will vary. In Trial 1, volume setting 8 corresponds to a sound level of 86 dB.

3. Answers will vary. At a sound level of 86 dB, the maximum exposure per day should not exceed 8 h.

4. Answers will vary depending on the sound level that each portable radio or tape player produces. Both tape players in the sample are potentially dangerous, since they can produce sound levels above 90 dB.

Extension and Application

1. Jack needs a sound level 5 dB greater than normal, which is three times as intense. Caroline needs a sound level 25 dB greater than normal to detect sound, which is approximately 300 times as intense.

2. $(10 \text{ dB}) \log \dfrac{I_2}{I_1} = \beta_2 - \beta_1$

$$\log \dfrac{I_2}{I_1} = 50 \text{ dB}/10 \text{ dB}$$

$$\dfrac{I_2}{I_1} = 100{,}000$$

The rock group makes 100 000 times as much noise as does normal conversation at 1 m.

3. In an enclosed room, sound is reflected from the ceiling, walls, floor, and furniture. It doesn't weaken as quickly and may be focused or amplified.

4. a. $I = \dfrac{100 \text{ W}}{4\pi \, (2.5 \text{ m})^2} = 1.27 \text{ W/m}^2$

b. $\beta = (10 \text{ dB}) \log \dfrac{1.27 \text{ W/m}^2}{10^{-12} \text{ W/m}^2} = 121 \text{ dB}$

The sound level that can be generated by a 150-W amplifier is just above the threshold of pain and could cause severe hearing damage. This hazard is compounded because the system will be installed in a small room where sounds reflect off walls, ceiling, etc., rather than dissipate in open space. To save her hearing, the student should not buy this powerful amplifier.

5. While the student has doubled the intensity of the sound, the increase in the sound level is 3 dB. The new sound level is 103 dB.

15-2

Resonance in an Open Tube

Introduction

Resonance forms in an open tube when constructive interference increases pressure, particle displacement, and sound amplitude. Adjusting the length of an open tube reinforces the incident wave, increasing the amplitude of the vibrations. The sound wavelength is directly proportional to the resonant length of the tube and its diameter. Using the frequency of the sound source and the wavelength, the wave formula, $v = f\lambda$, gives the speed of sound. The experimental results will be affected by temperature and humidity—sound travels faster as these increase.

Process Skills

observing and inferring; using numbers; measuring in SI; interpreting data; hypothesizing; experimenting; using variables, constants, and controls

Preparation Hints

- Small cardboard tubes from rolls of toilet tissue, paper towels, and gift wrap work well. A selection of different tubes should provide several sets that slide into each other and adjust to different lengths.

- Using the information from Figure A, you can prepare a variety of tubes that are close to the length needed to produce resonance.

For example, a tube adjustable from 25 to 45 cm can be used with tuning forks of frequencies between 512 and 384 Hz.

Materials

adjustable open-pipe tube, 0.25–1.4 m long
tuning-fork hammer
2 tuning forks, with frequencies between 126 Hz and 512 Hz
thermometer
meterstick

Procedure Hints

The resonance in an open tube is not as loud as resonance in a closed tube. However, it is still easy to adjust the tube length to produce the maximum sound. Cardboard tubes do not produce sounds as loud as those generated in a smooth tube, like the glass closed tubes.

Data and Observations

Table 1 (Sample data are given.)

Trial	Frequency (Hz)	Tube diameter (m)	Measured length of tube (m)
1	256	0.0480	0.630
2	384	0.0480	0.415

Table 2 (Sample data are given.)

Trial	Corrected length (L) of air column (m)	Wavelength $\lambda = 2L$ (m)
1	0.668	1.34
2	0.453	0.906

Table 3 (Sample data are given.)

Trial	Air Temperature (°C)	Velocity correction for temperature (m/s)	Accepted velocity of sound (m/s)
1	21	12.4	344
2	21	12.4	344

Analysis and Conclusions

1. $v_1 = f\lambda = (256 \text{ Hz})(1.34 \text{ m}) = 343 \text{ m/s}$
$v_2 = f\lambda = (384 \text{ Hz})(0.906 \text{ m}) = 348 \text{ m/s}$

2. $\text{Error}_1 = (344 - 343)(100\%)/344 = 0.3\%$
$\text{Error}_2 = (348 - 344)(100\%)/344 = 1.2\%$
Students should have relative errors of less

than 10%.

3. The steam will change the velocity of sound because of the greater temperature and density of steam compared with those of air. The velocity will increase. (The velocity of sound is about 25% faster in steam than in air.)

4. Resonance should recur when the tone's wavelength is about four times the length of the open tube. Students should find a tube of approximately twice the length that produced resonance for a given tuning fork and repeat the procedure. The second harmonic is not as loud as the fundamental resonant sound.

Extension and Application

1. The frequency of the note produced is 55 Hz.
$$v = 331.5 \text{ m/s} + (0.59 \text{ m/s/°C})(20°C)$$
$$= 343.3 \text{ m/s}$$
$$\lambda = 2(l + 0.8d) = 2[3.0 \text{ m} + (0.8)(0.15\text{m})]$$
$$= 6.24 \text{ m}$$
Thus, $f = v/\lambda = (343.3 \text{ m/s})/(6.24 \text{ m})$
$$= 55 \text{ Hz}$$

2. $d = vt = 343 \text{ m/s} \times 4 \text{ s} = 1372 \text{ m}$, or nearly a mile

16-1

Polarized Light

Introduction

The process of light polarization causes the light waves to vibrate in only one plane. Polarizing filters can be employed to produce this effect. The filters can be used in combination and rotated through various angles. The changes in the intensity of the light transmitted at various angles of filter rotation can then be noted. Such rotation can also reveal patterns that are useful in analyzing stresses placed on various materials.

Different objects differ in their tendency to polarize light that is reflected from them. Some objects, such as glass mirrors, do not polarize the reflected light. Others, such as transparent glass plates and water, do produce polarization. Polarizing filters can also be used to study such differing effects of light reflection.

Process Skills

observing and inferring, classifying, analyzing, interpreting data, hypothesizing

Preparation Hints

- Special elastic-stress, optical plastic can be purchased from science supply houses. It can be used in step 8.

- Small plastic boxes or protractors, which have been stamped with holes or notches, are excellent materials for observation of stress.

Materials

2 polarizing filters, disks, or sheets
Iceland spar calcite crystal
clear plastic object, such as a box, protractor, or Plexiglas rod
lightbulb, 40 W, clear filament, with socket and cord
plane mirror
safety goggles
glass plate
protractor (optional)
ruler (optional)

Procedure Hints

The extension could be delayed until refraction is more fully explained in Chapter 17.

Data and Observations

Table 1
Observations of light source with 1 polarizing filter:
The light intensity does not change.
Observations of light source with 2 polarizing filters, rotating the filter closer to the light:
During one rotation, the intensity changes twice, from minimum to maximum intensity.
Observations of light source with 2 polarizing filters, rotating the filter closer to your eye:
The same variation as above was observed.

Table 2
Drawing of a word viewed through Iceland spar:
The word appears twice, and the images are side by side and slightly off center. If the crystal is rotated, one of the words moves around the other, transcribing a circle around the stationary word.
Observations of a word viewed through Iceland spar, with 1 rotating polarizing filter:
As the filter rotates, one image disappears and then reappears as the other image fades away.

Table 3
Observations of light reflected from plane mirror:
The light intensity does not change.
Observations of light reflected from glass plate:
The light intensity changes.
Observations of light reflected from shiny tabletop:
The light intensity changes.

Table 4
Observations of plastic:
A pattern of color appears in the plastic. The colors change quickly and move together under stress. Where the strain is greatest, the colors are closest.

Analysis and Conclusions

1. Light from a lightbulb is not polarized; the intensity does not change as the filter rotates.

2. The intensity changes twice from minimum to maximum brightness as one of the filters rotates through 360°. It makes no difference which filter rotates. The maximum change occurs when one filter rotates 90°.

3. The light images were polarized, and one image shifted away from the other. The light rays of the images travel at different speeds, so they bend (refract) at different angles. (Materials like Iceland spar are called double refracting, because they have two indices of refraction.)

4. Light reflected from a plane mirror is not polarized. Its intensity does not change when observed through one rotating polarizing filter. If students are using a thick glass mirror, they may observe some polarization.

5. Light reflected from a glass plate or a shiny surface is polarized; the light changes intensity when viewed through one rotating polarizing filter.

6. Color bands, some wide and some narrow and closely spaced appeared. The bands moved together as the material was squeezed.

Extension and Application

1. A piece of polarizing material oriented in one plane, such as vertically, could cover the high-beam lights, while polarizing material oriented in the opposite direction, horizontally, could be in the windshield. Glare from headlights would be reduced. The drawback to this scheme is the need to outfit all cars with these materials in these orientations. Also, the filter material does block some light, which is not desirable at night.

2. Engineers could build a transparent plastic model of the product and subject it to various stresses and strains. When viewed through polarized light, the model would show the locations of greatest stress. Thus, the engineers would know where to put reinforcements or whether a new design is required.

3. Students should measure angles of 55°–60° for plate glass. This corresponds to an index of refraction of about 1.5. The index of refraction of water is 1.33.

16-2

Intense Light

Introduction

Students will devise a procedure for relating the power rating of light sources to their brightness. They will gather data on the wattage and light intensity of a colored lightbulb and several white lightbulbs, using a light probe and computer. They will use the data to graph the data and discover the relationship between wattage and intensity. In this lab, they will apply the concept of luminous intensity, discussed in Chapter 16 of their textbook, and of power, discussed in Chapter 10.

Luminous intensity is measured in candelas, and the illuminance on a spherical surface a certain distance away is measured in lux. The probes in this lab measure brightness on a different scale. Students need not be concerned about the particular units because all the comparisons will be done on the same scale.

Process Skills

hypothesizing; using variables, constants, and controls; measuring; collecting and organizing data; making and using graphs; inferring

Possible Hypotheses

The hypothesis that should be supported by the results is that intensity increases with wattage, and that a colored lightbulb produces lower-intensity light than does a white bulb of the same wattage. Students might also hypothesize that the intensity does not vary with wattage, that it varies inversely, etc., and that the intensity of a colored bulb has another relationship to that of a white bulb of the same wattage. Such hypotheses should not be supported by the results.

Possible Materials

lamp
4 white light bulbs of different wattages
colored light bulb
heat-resistant gloves
CBL unit with light probe
TI-82 or TI-83 calculator with unit-to-unit link cable
L06LIGHT program for calculator
ring stand and clamp
books for stacking
light shield, such as curtains or cardboard

Possible Procedures

One procedure that works well is to set up a light probe connected to a CBL. With the unit-to-unit link cable, connect the CBL to a TI-82 or TI-83 calculator, using the I/O ports on the bottom edge of each unit. Connect the end of the light probe to channel 1 (CH1) on the top edge of the CBL. Turn on the CBL and the calculator. Enter the L06LIGHT program in the calculator.

Put on cotton gloves to put the 40-W bulb in the lamp. Attach the probe to a ring stand or put it on a pile of books, so that it is at the same height as the widest part of the bulb. Set the lamp 0.5 m from the probe. Record the bulb's wattage in a table. Turn on the light and read the intensity of the light. Record the results of three tests of the bulb's intensity. Turn off the lamp. Calculate the average intensity for the bulb. When the lightbulb is cool, remove it from the lamp and replace it with the 60-W bulb. Repeat the procedure for each white bulb and for the colored bulb, keeping the same distance between probe and source.

Students might test the bulbs in a different order or use a noncomputerized way of estimating brightness. However, the unaided eye is unlikely to produce quantitative results that are adequately objective.

Teaching Suggestions

- Have students work in pairs or in groups of three or four. Each student should collect some data. Be sure that students understand the operation of the light probe and CBL. Limit students to a probe-to-source distance that is close to 0.5 m.

- New lightbulbs will show more reliable relationships between actual wattages and intensities. Limit the range to 40–100 W, which will demonstrate a fairly linear relationship. The lightbulbs must be the same shape and kind.

- Remind students not to look at lit bulbs and to let the lightbulbs cool before handling. They should wear cotton gloves and goggles.

- Darken the room, leaving just enough light to carry out the procedure safely. Each group needs a clear area that is shielded from the lamps of other groups. The shields could be large boxes with one side removed, cardboard, or curtains. The light shields should extend at least 0.25 m above the lights.

- If you can, use black bulbs with wattage between 40 W and 90 W. Warn students that these bulbs become extremely hot and take a long time to cool. Lightbulbs of pale colors, such as yellow bug lights, do not offer much intensity contrast with white lightbulbs.

- L06LIGHT is the program for this lab. Enter in order:

```
PROGRAM:LO6LIGHT
:Lbl 0
:ClrHome
:Menu("*MAIN MENU*"),"RUN
ACTIVITY",A,"QUIT",Z)
:Lbl A
:PlotsOff
:FnOff
{1,0}→L₁
:Send(L₁)
:ClrHome
:{1,1,1}→L₁
:Send(L₁)
:ClrHome
:Disp "PRESS ENTER"
:Disp "TO ZERO
:Pause
:{3,.01,5,0}→L₁
:Send(L₁)
:ClrHome
:Disp "PRESS ENTER TO"
:Disp "READ INTENSITY."
:Pause
:{3,.01,5,0}→L₁
:Send(L₁)
:Goto 0
:Stop
:Lbl Z
:ClrHome
```

- Review students' plans, and accept reasonable procedures. Also review graphing techniques, and make sure that the concepts involved are clear.

Data and Observations

Distance between
light source and sensor: _____0.500_____ m

Table 1 (Sample data are given.)

Bulb Color	Wattage	Intensities			
		Trial 1	Trial 2	Trial 3	Average
white	40	0.836	0.840	0.838	0.838
white	60	0.851	0.849	0.847	0.849
white	75	0.858	0.855	0.855	0.856
white	90	0.860	0.863	0.857	0.860
black	75	0.832	0.830	0.828	0.830

Analyze and Conclude

1. Answers may vary. Sample answer:
   y-intercept = 0.820

2. $m = \dfrac{y^2 - y^1}{x^2 - x^1}$

   $= \dfrac{0.860 - 0.838}{90 - 40} = 4.0 \times 10^{-5}$

3. The point lies below the line. For the sample data, the vertical difference is approximately 0.026. This distance is the difference between the intensity of the colored bulb and that of a white bulb of equal wattage.

4. Answers may vary. The data point should be $x = 40$, $y =$ a value above the line for white bulbs. A high-intensity bulb should produce more light per watt than a comparable white bulb.

5. The graph would resemble that for the white bulbs, with a similar slope but at lower intensities.

6. The intensity would decrease. The light would illuminate a larger area, decreasing its brightness at a given point.

7. The intensity should drop to zero if the wattage became zero. The actual complete graph of intensity versus wattage is probably a curve. The useful range for the linear equation is roughly 35–100 W.

Apply

1. Answers may vary. The experiment could be a simple variation on the techniques used in this lab. For example, measure the intensity of each colored bulb using a probe, with distance and wattage kept constant. The intensities will vary , based on the quantity of light released from each bulb. In general, the darker the color of the bulb, the lower the intensity would be.

17-1

Reflection of Light

Introduction

A light ray striking a reflecting surface always reflects from that surface at an angle that equals the angle at which it arrived. The two angles are measured relative to a line perpendicular to the surface, the normal. The image of an object from a flat or plane mirror is always a virtual image, as it cannot be captured on a viewing screen. By tracing the light rays from an object reflected in a plane mirror, the apparent location of this virtual image can be found. Virtual images from plane mirrors are always the same size as the object and are right side up, but are reversed right to left. The distance from a mirror to the virtual image always equals the distance from the mirror to the object.

Process Skills

observing and inferring, recognizing spatial relationships, interpreting scientific illustrations, measuring in SI, using numbers, comparing and contrasting, communicating, collecting and organizing data

Preparation Hints

One setup for Part C can be used for all students. As they work on parts A and B, each lab group can take a turn with the laser. In this way, the laser remains in a separate, protected area of the room.

Materials

thin plane mirror
small wood block
metric ruler
protractor
2 sheets plain white paper
2 straight pins or dissecting pins
rubber band
corkboard or cardboard
thick plane mirror
4 thumbtacks or pieces of masking tape
He-Ne laser
glass plate (optional)
small blocks of wood (optional)

Procedure Hints

If students have difficulty constructing their triangles in Part B, tell them to repeat the procedure, but emphasize that they must sight very carefully along the lines to the image of the pin in the mirror.

Data and Observations

Table 1 (Sample data are given.)	
Object distance	4.1 cm
Image distance	3.8 cm
Angle of incidence, i_1, *PXN*	13°
Angle of reflection, r_1, *AXN*	13.5°
Angle of incidence, i_2, *PYN'*	24°
Angle of reflection, r_2, *BYN'*	25°

Table 2 (Sample data are given.)	
Points	Distance to mirror line (cm)
A	1.7
B	5.4
C	1.8
A'	1.8
B'	5.2
C'	1.9

Table 3
Drawing of reflected laser light from a thin plane mirror: A single large spot should appear.
Drawing of reflected laser light from a thick plane mirror: ● ⬤ ● ● • ·
Observation of reflected laser beams: A series of reflected beams becomes progressively narrower as they are farther from the incident beam.

Analysis and Conclusions

1. They are equal (or should be nearly so).

2. The image is the same distance behind the mirror as the object is in front of the mirror.

3. The image should be the same size as the object. Students may have some small errors due to sighting inaccuracies.

4. Images formed by plane mirrors are as far behind the mirror as the object is in front of the mirror. The image is reversed right to left and is the same size as the object.

5. The image appears where the light rays apparently cross, not where they actually cross. This image does not exist behind the mirror, where it appears to be.

6. The second spot is brightest, because this beam is the reflection from the mirrored surface. The first spot is the reflection from the glass.

7. The thin mirror reflects light from the glass and mirrored surface, but the reflected spots are nearly in the same location. The thick mirror has multiple reflections, spaced farther apart, from the glass at the top and the silver at the bottom of the mirror, as shown below.

Extension and Application

1. The distances should be equal.

2. Students should be unable to see an image of the card in the single mirror. The image of the card in the two mirrors is right side up and reads correctly right to left, as if you were looking directly at the card. Explanations will vary. The reflection from each mirror is flipped right to left by the other mirror. The image from the pair of mirrors is a virtual image. (A ray diagram would be rather complex to construct but might be a project for gifted students.)

3. Some of the light is transmitted while most light is reflected. One-way mirrors reflect a large portion of the incident light, such as 50%–60%, while allowing a smaller portion, such as 5%–15%, to pass through. Since a smaller portion of the light is transmitted, the viewing area is usually darker than the observed area.

17-2

Snell's Law

Introduction

The speed of light varies depending on the medium through which it is traveling. The ratio of the speed of light in a particular medium and the speed of light in a vacuum is the index of refraction for that medium. As the light passes from one medium to another, it refracts at the boundary between the different media. The index of refraction for a medium may be calculated from the measurement of the angle at which a light ray refracts when it passes from one medium to another. Snell's law can be used to calculate the index of refraction for a medium if the angle of incidence and the angle of refraction are known. This experiment uses Snell's law to calculate the index of refraction for a piece of glass.

Process Skills

observing and inferring, recognizing spatial relationships, interpreting scientific illustrations, measuring in SI, using numbers, comparing and contrasting

Preparation Hints

An alternative to plate glass might be thick plastic sheets. The refractive index of most clear plastic ranges from 1.45 to 1.55.

Materials

glass plate
metric ruler
protractor
plain white paper
small samples of glass, garnet, tourmaline, beryl, topaz, and quartz (optional)
100-mL bottle of mineral oil or oil of wintergreen (optional)
100-mL bottle of clove oil (optional)
100-mL bottle of water (optional)
100-mL bottle of olive oil (optional)
100-mL bottle of cinnamon oil (optional)
tweezers (optional)

Procedure Hints

- This activity should be easy to understand and complete, even for students who have not studied trigonometry. However, review the relationships of triangles prior to the lab and introduce Snell's law as a relationship between the opposite side and the hypotenuse of a triangle.

- A dish of water or oil could be substituted for the glass plate to provide a different index of refraction. However, the oil is messy if it is spilled.

Data and Observations

Table 1 (Sample data are given.)					
θ_i	θ_r	$\sin \theta_i$	$\sin \theta_r$	$\theta_{r'}$	Index of refraction, n
30°	19°	0.5	0.326	30°	1.53
45°					

Analysis and Conclusions

1. The indices should be equal or nearly equal. Students might have as much as 10 percent error if they are not careful while sighting along their rulers.

2. As the light rays pass at an angle from a less dense to a more dense medium, they bend toward the normal.

3. As the light rays pass at an angle from a more dense to a less dense medium, they bend away from the normal.

4. The two angles are equal. According to Snell's law, as the light ray passes back to the optically less dense medium, it should resume its original speed and bend away from the normal.

5. $v_s = c/n_s = (3.00 \times 10^8 \text{ m/s})/1.53$
   $= 1.96 \times 10^8 \text{ m/s}$
   Light travels 3.00×10^8 m/s $- 1.96 \times 10^8$ m/s $= 1.04 \times 10^8$ m/s faster in a vacuum.
   % increase $= 1.04 \times 10^8/1.96 \times 10^8 = 53\%$

Extension and Application

1. Light will bend more traveling from air into water than from water into glass. Using the general form of Snell's law, $n_i \sin \theta_i = n_r \sin \theta_r$, light traveling from air into water yields an effective n of 1.34, while light traveling from water into glass yields an effective n of 1.14, determined from $(1.34)(\sin \theta_i) = (1.53)(\sin \theta_r)$. The greater the index of refraction, the greater the amount of bending.

2. The sample materials should be easy to identify. The samples will be just barely visible in the liquid with nearly the same index of refraction. Glass normally has an index of refraction of about 1.5. You probably will not have any glass with $n = 1.7$.

18-1

Concave and Convex Mirrors

Introduction

A mirror that is a section of a sphere is a spherical mirror. The center of the sphere is the center of curvature, and the focal point is located halfway between this point and the mirror surface. The distance from the focal point to the mirror surface is the focal length of the mirror. The image formed by a concave mirror is a real image if the object is farther from the mirror than the focal length. When the object is at the focal point, no image forms, and when the object is between the focal point and the mirror, a virtual image results. The real images formed by a concave mirror are always between the mirror and the object. The image from a convex mirror is always a virtual image and is behind the mirror.

Process Skills

using numbers, measuring in SI, interpreting scientific illustrations, observing, communicating, formulating models, interpreting data, experimenting

Materials

concave mirror
convex mirror
2 metersticks
meterstick support
cardboard screen
holders for mirror, screen, and light source
masking tape
metric ruler
light source

Preparation Hints

- Object markers may be used with the light sources. However, students usually have no trouble identifying the orientations of the images. Lightbulbs generally have flat bases or glowing filaments, which indicate upright or inverted positions.

- Candles can be substituted for light sources. If candles are used, caution students about the flames.

- Students will have less difficulty locating sharp images if the mirrors are good. To check their quality, take clean mirrors and focus sunlight onto a screen. Good mirrors will produce a small image of the sun, while poor mirrors will not focus the sunlight.

Procedure Hints

- Allow ample time to complete this investigation. A prelab discussion of the types of images formed by mirrors and a demonstration of how to find focal lengths may be helpful.

- Use a concave mirror to project an outdoor scene from outside a window onto a screen. This demonstration will familiarize students with mirrors and real images, and the color image is interesting.

- The lab must be dark, so that images are easy to see. The electric light sources should give enough light while students set up equipment. One window should be unshaded for Part A.

- Tell students to work quickly when using the sun as an image source for Part A. The focused image could burn the screen.

Data and Observations

Table 1 (Sample data are given.)	
Focal length of mirror, f	33 cm
Center of curvature of mirror, C	66 cm
Height of light source, h_o	1.0 cm

Table 2 (Sample data are given.)

Position of object	Beyond C	At C	Between C and F	At F	Between F and A
d_o	1.0 m	66 cm	48 cm	33 cm	16 cm
d_i	49 cm	66 cm	98 cm	—	—
h_i	0.8 cm	1.0 cm	6 cm	—	larger than object
Type of image	real	real	real	none	virtual
Direction of image	inverted	inverted	inverted	no image	erect

Table 3 (Sample data are given.)

Trial	Position of object	Position of image	Type of image	Image size compared to object size	Direction of image
all	all positions	behind mirror	virtual	smaller	erect

Analysis and Conclusions

1. **a.** The image is between C and F, inverted, real, and smaller than the object.

   b. The image is at C, real, inverted, and the same size as the object.

   c. The image is beyond C, real, inverted, and larger than the object.

   d. No image is formed.

   e. The image is virtual, erect, and larger than the object.

2. The images are virtual, smaller and erect, and behind the mirror.

3. Beyond C, $f = 33$ cm; at C, $f = 33$ cm; between C and F, $f = 32$ cm. Student values should be in good agreement.

4. The average is 33 cm, relative error is 0%. Students should have an error of no more than 10%–15% if they have made careful observations and have reasonably good mirrors.

Extension and Application

1. The top edge of the box should be along the principal axis, while the lightbulb socket should be at C to produce an image equal in size to the object and inverted.

18-2

Convex and Concave Lenses

Introduction

Lenses, both concave and convex, form images in a manner that is similar to the image formation of mirrors. Concave lenses can form only virtual images because they cause all light rays to diverge. Convex lenses can form real images, virtual images, or no image. If an object is farther from the lens than twice the focal length of the lens, then the image is inverted, real, and smaller than the object. If an object is between the focal length and twice the focal length, the image is erect, real, and enlarged. An object at the focal point forms no image with a convex lens, because all the light rays exiting the lens are parallel to each other. If an object is between the lens and the focal point, a virtual image forms.

Process Skills

using numbers, measuring in SI, observing, communicating, formulating models, making and using graphs, interpreting scientific illustrations, interpreting data, experimenting

Preparation Hints

• Check the lenses to make sure they are clean and free from scratches. Dirty or scratched lenses will impair image formation and make it hard for students to focus an image.

- When using sunlight to determine the focal length of the concave lens, the contrast between the circle of light projected on the screen and the ambient light can be increased by surrounding the lens with a piece of paper so that sunlight can't fall directly on the screen.

Materials

double convex lens
concave lens
meterstick
meterstick support
small cardboard screen
light source
holders for lens, screen, and light source
metric ruler
sunlight or HeNe laser

Procedure Hints

- The lab must be dark, so that images are easy to see. The electric light sources should give enough light while students set up equipment. One window should be unshaded for Part A.

- If the sun is the object, remind students to measure the distance quickly because a focused spot can become hot and burn the screen. Caution students to avoid looking directly at the sun.

Data and Observations

Table 1 (Sample data are given.)

Focal length, f	18	cm
$2F$	36	cm
Height of light source, h_o	1.2 cm	

Table 2 (Sample data are given.)

Position of object	Beyond 2F (cm)	At 2F (cm)	Between 2F and F (cm)	At F (cm)	Between F and lens (cm)
d_o	50	37	25	17.5	10
d_i	29.2	37.2	66.5	–	–
h_i	1.0	1.2	1.8	–	appears larger through lens
Type of image	real	real	real	none	virtual
Direction of image	inverted	inverted	inverted	–	erect

Table 3

Distance from lens (m)	Diameter of screen image (cm)
0.0	0.15
0.5	0.5
1.0	0.75
1.5	1.1
2.0	1.4
2.5	1.8
3.0	2.0

Analysis and Conclusions

1. **a.** The image is between F and $2F$, real, about the same size, and inverted.

   b. The image is at $2F$, real, the same size as the object, and inverted.

   c. The image is beyond $2F$, real, smaller, and inverted.

   d. No image is readily apparent.

   e. The image is virtual, larger, erect, and on the same side of the lens as the object.

2. Beyond 2F, f = 18.4 cm; at 2F, f = 18.5 cm; between 2F, and F, f = 18.2 cm There should be good agreement for all measurements.

3. Average f = 18.4 cm; relative error = (18.4 − 18)(100%)/18 = 2%. Students should have a relative error of less than 10%.

4. The intersection provides a focal length of −0.25 m. The accepted value is −0.20 m, therefore the relative error = (0.25 − 0.20)(100%)/0.20 = 25%. Student error may vary greatly because the observed diameters are usually not clearly defined at the edges.

Image Size vs Distance

This graph was produced with the Lab Partner graphing program. The x-intercept is −0.26 m.

Extension and Application

1. The object looks normal when viewed through a convex or a concave air lens in air. However, the object looks smaller when viewed through the convex air lens immersed in water and larger when viewed through the concave air lens immersed in water. Note the reversal of the usual effects of glass lenses on these shapes. In water, light rays are traveling from the object through a more dense medium (water) to a less dense medium (air in the lenses). This reversal of media accounts for the opposite effects of the lenses.

2. If the direction of the reflected laser pattern is opposite that of the observer's head, the person is nearsighted. If the direction of the reflected laser pattern is the same as that of the observer's head, the person is farsighted. If the observer does not perceive any relative motion, his or her eyesight is normal. To simulate nearsightedness or farsightedness, students could hold various lenses in front of their eyes and observe the laser light spot.

19-1

Double-Slit Interference

Introduction

This investigation explores the interference of light waves. The diffraction of light waves passing through two closely spaced slits shows alternating bright and dark bands, sometimes referred to as lines. Constructive interference produces the bright bands, and destructive interference produces the dark bands. The effect is most easily observed when light waves of a single wavelength pass through the slits. The pattern of the alternating bands is related to the spacing of the slits, the distance from the slits to the observing screen, and the wavelength of the light passing through the slits. The wavelength of the light can be determined from measurements of these distances.

Process Skills

measuring in SI, observing and inferring, using numbers, interpreting data

Preparation Hints

Double slits can be purchased or constructed. If you want students to construct their own double slits, tape a pair of razor blades together and lightly pull the unit across a microscope slide covered with dried graphite emulsion. An alternative to graphite emulsion is aluminum foil, placed over a microscope slide and cut with the pair of razor blades.

If you have dichroic filters available, these may give more easily distinguished lines, since the light they pass is almost monochromatic. These filters are usually rigid, however, and must be placed directly in front of the slits.

Materials

double-slit film
clear filament light (40 W) with socket
red, yellow, and violet transparent filters
rubber band
meterstick
metric ruler
magnifier with scale
masking tape
index cards, 12 cm × 18 cm
He-Ne laser or laser pointer (optional)

Procedure Hints

- Suggest that students use the red filter first, since it provides the best observations of diffraction and interference.

- If students have difficulty putting the filters on the lightbulbs, they can try holding the filters in front of the slits.

- If a variety of color filters is available, give each lab partnership a red piece and two other colors, so that the other colors are different for different lab partners. This individualizes the lab, and partners are less inclined to copy data.

- If students have difficulty measuring x for the violet filter, suggest that they move farther away (increase L). This color is the most difficult to observe and measure, especially if the room lights are on.

Data and Observations

Table 1 (Sample data are given.)	
	Observations of Light
Close	The bright bands of red are close together.
Far away	The bright bands of red move farther apart and are easier to see.

Table 2 (Sample data are given.)			
Color	**d (mm)**	**x (m)**	**L (m)**
Red	0.15	0.017	4.0
Yellow	0.15	0.015	4.0
Violet	0.15	0.012	4.0

Analysis and Conclusions

1. The pattern spreads out, so the lines are farther apart.

2. $\lambda_{red} = xd/L$
   $= (0.017 \text{ m})(0.15 \times 10^{-3} \text{ m})/4.0 \text{ m}$
   $= 640 \text{ nm}$
   $\lambda_{yellow} = xd/L$
   $= (0.015 \text{ m})(0.15 \times 10^{-3} \text{ m})/4.0 \text{ m}$
   $= 560 \text{ nm}$
   $\lambda_{violet} = xd/L$
   $= (0.012 \text{ m})(0.15 \times 10^{-3} \text{ m})/4.0 \text{ m}$
   $= 450 \text{ nm}$

   Answers will vary and will depend on how carefully students measure x. Answers should be within the visible spectrum (in the range of 400–700 nm).

3. Red light is diffracted most. Using the same values for L and d, the angle will be larger, since x is greater for the larger wavelength.

4. Answers will vary. The distance should be greater than the measured distance of x for the red light. Infrared light will be diffracted more than visible red light, since it has a larger wavelength.

5. Answers will vary. The distance should be less than the measured distance for x for the violet light. Ultraviolet light will be diffracted less than visible violet light, since it has a smaller wavelength.

Extension and Application

1. Answers will vary, depending on the values for L and x. However, when students substitute their values into $d = \lambda L/x$, the resulting value for d should be very close to the actual value.

2. Since $\tan \theta = x/L$ and $\theta = 0.8°$, $\lambda = xd/L = d(\tan \theta) = (0.05 \times 10^{-3} \text{ m})(\tan 0.8) = 700$ nm, which is the wavelength of red light.

3. $L = xd/\lambda$

$$= \frac{(14.7 \times 10^{-3} \text{ m})(1.50 \times 10^{-5} \text{ m})}{5.00 \times 10^{-7} \text{ m}}$$

$$= 0.441 \text{ m}$$

4. $x = L\lambda/d$

$$= \frac{(4.41 \times 10^{-1} \text{ m})(5.89 \times 10^{-7} \text{ m})}{1.50 \times 10^{-5} \text{ m}}$$

$$= 0.0173 \text{ m}$$

19-2

White-Light Holograms

Introduction

Laser light can be used to produce an image, called a hologram, that has a three-dimensional appearance. The hologram is a record of the constructive and destructive interference patterns that occur when the hologram is made. Since the hologram is a record of the interference patterns of light waves, it can show details as small as one-half the wavelength of the light used to create the hologram. Holograms are useful as tools to examine very small vibrations in objects and to measure very small details in objects. A laser light source is necessary to make holograms because laser light has a single wavelength and all the waves are in phase. A hologram that can be viewed with an ordinary light source rather than requiring a laser light source is called a white-light, or reflection hologram. In this experiment, students will produce a hologram and

examine its properties.

Process Skills

observing and inferring, comparing and contrasting, interpreting data, communicating

Preparation Hints

- Small plastic microwave trays can be substituted for the glass or plastic trays. Keep the trays as close to the size of the film as possible to minimize waste of chemicals.

- A massive object, such as an oscilloscope placed on its side, can be substituted for the steel plate.

- A small steel plate can be taped on any large nonsteel object and used as the base plate for the experiment. The magnets stick easily to this surface.

- A photographic darkroom timer with luminous hands would be an excellent alternative to the stopwatch.

Materials

He-Ne laser
small concave mirror or beam-diverging lens
  (expander)
holographic film plates (or holographic film and
  two glass plates)
automobile inner tube
bicycle air pump
2 index cards, 12 cm $\times$ 18 cm
holographic developer and bleach solution
plastic tongs
rubber gloves
green safelight
3 plastic trays, 10 cm $\times$ 10 cm
quarter or other shiny object to photograph
steel plate, 30 cm $\times$ 30 cm
helping-hand project holder
12 disk magnets
stopwatch or watch with second hand
black paper or posterboard
front-surfaced mirror, 10 cm $\times$ 15 cm (optional)
notebook-paper clips (optional)

Procedure Hints

- Provide one classroom setup of the laser and photographic platform. A small night-light with a green Christmas lightbulb works well as a safelight.

- For purposes of supervising chemical use, it

is recommended that the teacher develop the holograms at a central location. The teacher should wear a protective apron, goggles, and rubber gloves. Follow the manufacturer's directions for mixing and using the photographic chemicals. Lab partners can arrange their materials, take their picture, and bring the hologram to the teacher for developing. After it is developed, return the hologram to the students for viewing. Use the plastic tongs to transfer the plate from the developer to the bleach and to the water rinse.

- A diverging lens used as a beam expander can be substituted for the small concave lens. In this case, align the glass plate vertically, rather than horizontally. Place the object to be photographed behind the glass plate.

- You may need exposure times as long as 30 s to get a good image; however, if your laser is close and more powerful than 1.5 mW, shorter exposure times will work.

Data and Observations

Table 1 (Sample data are given.)	
	Observations of Holograms
Looking through hologram	The plate is transparent and greenish. No image is visible.
Looking at reflection	The image of the photographed objects appears in great detail. The objects have three dimensions, and the sides of the objects can be viewed as the hologram is tilted back and forth.

Analysis and Conclusions

1. The hologram appeared only in the reflected light and was not visible in the transmitted light. The reflected light recreated the interference pattern that produced the original hologram, thus making the image visible.

2. The hologram is three-dimensional and has depth; it is not two-dimensional, as is a picture or photograph. The hologram does not appear as sharp or as clear as the photograph.

3. The color comes from the interference pattern resulting from the white light, not from anything on the film.

Extension and Application

1. The transmission hologram is of much better quality and appears sharper than the white-light reflection hologram. However, it must be viewed in laser light.

2. Holograms are used to reduce or prevent counterfeiting of money and credit cards. Holograms of rare coins produce a three-dimensional record of the coin collection and can be used for positive identification.

20-1

Investigating Static Electricity

Introduction

Static-electric charges appear on the surface of nonconducting materials when those materials have an excess or deficiency of electrons. Static charges are often generated by contact between materials. Electrons are removed from one material, giving it a positive charge, while the other material gains those electrons and becomes negatively charged. A conductor will maintain a static charge only when the charged object is ungrounded. Two charges of equal magnitude and opposite polarity will neutralize each other when they touch.

A charged object can produce a charge of opposite polarity in another object without coming into contact with that object. This process is induction. When an object with an induced charge touches the object that induced the charge, the object with the induced charge acquires a charge of the same polarity as the object that induced the charge. Objects with charges of opposite polarity attract each other, and objects with like charges repel each other. The electroscope uses the repulsion of like charges to indicate static charge.

Process Skills

using variables, constants, and controls; observing and inferring; comparing and contrasting; hypothesizing

Preparation Hints

- On a humid day, it may not be possible to produce a positive charge with the glass rods. In these weather conditions, the lucite or acrylic materials produce adequate positive charges. The strength of the positive charge produced by these materials can be verified by comparison with the negative charge produced by a hard-rubber rod.

- Clean any dust or dirt from the equipment used for this experiment, particularly the electroscope. Clean equipment will produce the best results, particularly in humid weather.

- Nylon or cotton scraps can also be used for rubbing pads.

Materials

hard-rubber rod or vinyl strip
glass rod
plastic wrap or acrylic rod
silk pad
wool pad
pith ball suspended from holder by silk thread
leaf or vane electroscope
Leyden jar
wire with alligator clips

Procedure Hints

- Encourage students to make detailed observations. Often students rush through this experiment and are unsure of what they saw.

- On a dry day, the vane electroscope should perform well. It requires more charge to cause a deflection than does the leaf electroscope. On more humid days, it is better to use a leaf electroscope.

Data and Observations

Table 1
Observations of pith ball with nearby charged rod:
The pith ball is attracted toward the charged rod.
Observations of pith ball after it has been touched with a charged rod:
The pith ball is attracted toward the rod, but after it has been touched, it is repelled.

Table 2
Observations of pith ball with nearby charged rod:
The pith ball is attracted toward the charged rod.
Observations of pith ball after it has been touched with a charged rod:
The pith ball is attracted toward the rod, but after it has been touched, it is repelled.

Table 3
Observations of uncharged electroscope when negatively charged rod touches it:
The vane deflects (or the leaves separate).
Observations of negatively charged electroscope when negatively charged rod is nearby:
The electroscope vane (or leaves) deflects more while the rod is nearby; it returns to the original charged position when the rod is removed.
Observations of negatively charged electroscope when a positively charged rod is nearby:
The vane moves toward zero (or the leaves collapse). When the rod is removed, the electro-scope again indicates charge.
Observations of deflection of leaves or vane when plastic material charges the electroscope and a negatively charged rod is nearby:
The rubber rod causes the electroscope to move toward zero, indicating that it has a charge opposite that of the rod. The electroscope is positively charged.

Table 4
Observations of electroscope when charged rod is nearby:
The electroscope vane or leaves move, indicating the presence of charge, but when the rod is removed, the vane returns to zero or the leaves fall.
Observations of electroscope after touching wih finger:
The electroscope indicates a charge when the charged rod is brought near, but drops to zero when it is touched by a finger. After the finger and then the charged rod are removed from the electroscope top, the electroscope again indicates a charge.
Observations to determine the type of charge:
The electroscope has a charge opposite that of the charged rod that was originally nearby. The leaves converge or the vane returns to zero, indicating that the original charged rod is attracting the charge of the electroscope.

Table 5
Observations of charging a Leyden jar–electroscope combination:
The electroscope charges easily by itself but does not (apparently) charge when the Leyden jar is attached. The charge travels through the wire to the Leyden jar, which stores charge.

Analysis and Conclusions

1. The ball touches the charged rod, becomes charged, and then is repelled.

2. The ball touches the charged rod, becomes charged, and then is repelled.

3. The process is the same for either type of charge.

4. When the charged rod touches the electroscope, the vane deflects (leaves diverge). The vane or leaves return to normal when the electroscope is touched by a finger.

5. Electrons were transferred to the electroscope, and the vane (leaves) became negatively charged and deflected (repelled each other).

6. The leaves were still negatively charged.

7. The charge was positive. Results may vary depending on the type of material available.

The charge type can be confirmed by testing it with a known negatively charged rod.

8. The finger provided a path to ground for the same charge as that of the charged rod brought near the electroscope. The net charge remaining on the electroscope, after the finger and rod were removed from the electroscope top, was opposite that of the charged rod.

9. The charge was opposite that of the rod.

10. It acted as a ground, removing some of the charges.

11. Conduction produces the same charge; induction produces the opposite charge.

12. The charge was transferred to the Leyden jar (it stores static charges) from the electroscope. The electroscope did not retain the charge, as shown by the action of the leaves (they came together) or the vane (it returned to zero).

Extension and Application

1. The particles gain a charge, by induction, that is opposite the charge on the plates. The particles, having an opposite charge, are then attracted to the plate. Once the particle has touched the plate, it usually remains attached because of unequal charge distribution over the surface of the particle. Students should guess that the electrostatic filter works because most dirt and dust particles are not good conductors of electricity. An electrostatic filter has the advantage that it can remove particles that are much smaller than would be practical using a mechanical type of filter.

21-1

The Capacitor

Introduction

The unit of capacitance is the farad. One farad is such a large quantity that practical devices have values measured in microfarads or even picofarads. The dielectric in large-value capacitors, like the ones used in this experiment, is usually a microscopically thin film of insulating oxide material on the surface of one or both of the capacitor's plates. Since capacitance is inversely proportional to the spacing of the plates in a capacitor, the extreme thinness of the dielectric gives a large capacitance for a small plate area. The electric charge flows into a charging capacitor or out of a discharging capacitor, not through the capacitor. Once a capacitor has reached equilibrium, the current flow is zero.

Misconception: Students may think that the voltage rating on a capacitor means that the capacitor has the marked capacitance only at that voltage, or that the capacitance of a capacitor depends on the applied voltage. Capacitance is independent of voltage. The voltage rating on a capacitor indicates the insulation ability of the dielectric material in the capacitor.

Process Skills

observing and inferring, measuring in SI, making and using graphs, using numbers, collecting and organizing data, interpreting data

Preparation Hints

- Larger-value capacitors may be substituted for the 1000-μF value; as the value increases, however, the capacitor takes longer to charge. If substantially smaller values are used, the capacitor will charge too quickly, making measurement difficult.

- Before students answer the questions, you might explain or demonstrate how to estimate the area under a curve and why this area divided by the series resistance represents the total charge on the capacitor.

Materials

CBL unit
voltage probe
link cable
graphing calculator with PHYSICS CBL software
   program
1000-μF capacitor
10-kΩ resistor
27-kΩ resistor
15-VDC source (battery or power supply)
SPDT knife switch
connecting wire

Procedure Hints

- Discuss polarity for meters, probes, batteries, and capacitors prior to the experiment. This preparation will minimize the chances that students will wire the circuit incorrectly.

Before they close the switch, have students trace the circuit path with a finger while looking at the circuit diagram.

- Show students how the unit of *RC* (seconds) can be traced from the derived SI units of resistance (ohms) and capacitance (farads).

Data and Observations

Capacitance: _____ 1000 μF _____

Table 1 (Sample data are given.)		
Time (s)	27 kΩ Voltage (V)	10 kΩ Voltage (V)
0	13.8	11.0
5	12.2	9.1
10	10.8	6.41
15	9.2	4.50
20	8.1	3.21
25	6.75	2.29
30	5.67	1.71
35	5.14	1.31
40	4.3	1.01
45	3.79	0.81
50	3.24	0.59
55	2.7	0.50
60	2.57	0.46
65	2.24	0.39
70	2.02	0.36
75	1.72	0.32
80	1.57	0.30
85	1.41	0.29
90	1.29	0.25

Analysis and Conclusions

1. It began fully charged to the battery potential difference and slowly discharged through the series resistor.

2. The resistor controls the amount of current flowing to the capacitor; thus, it regulates the charging rate. The larger the value of resistance, the longer it will take to charge the capacitor.

3.

Trial 1 Discharging Capacitor

Trial 2 Discharging Capacitor

These graphs were produced with the Lab Partner graphing program.

4. Charge = $\text{Area}_{27\ k\Omega}$ = $\frac{1}{2}bh$
$\qquad = \frac{1}{2}(90\ s)(10.8\ V)/(27\ 000\ \Omega)$
$\qquad = 1.8 \times 10^{-2}\ C$

Charge = $\text{Area}_{10\ k\Omega}$ = $\frac{1}{2}bh$
$\qquad = \frac{1}{2}(35\ s)(11.0\ V)/(10\ 000\ \Omega)$
$\qquad = 1.9 \times 10^{-2}\ C$

5. Trial 1: $C = q/V = (1.8 \times 10^{-2}\ C)/15.0\ V = 1.2 \times 10^{-3}\ F = 1200\ \mu F$

Trial 2: $C = q/V = (1.9 \times 10^{-2}\ C)/15.0\ V = 1.3 \times 10^{-3}\ F = 1300\ \mu F$

6. Trial 1: relative error = $(1200 - 1000)(100\%)/1000 = 20\%$

Trial 2: relative error = $(1300 - 1000)(100\%)/1000 = 30\%$

7. Without a resistor, the capacitor would become charged almost instantly; the current would quickly drop to zero.

8. The capacitor value and the power source voltage remained constant while the resistance was manipulated. The current in the circuit varies inversely with time.

Extension and Application

1. Answers will vary, depending on the capacitor used. The exponential decay of current shown on the graph will drop more slowly as the capacitance increases.

2. Since a capacitor requires a specific time to charge through a resistor, the RC combination can be used for timing, such as for a burglar alarm or a timer to turn on or off a light. With a short time of charging or discharging, the RC circuit could produce musical notes for an electronic instrument or keyboard.

22-1

Ohm's Law

Introduction

Ohm's law, $V = IR$, is a statement of the relationship between voltage, current, and resistance. Ohm's law is probably the most often used single equation in electronics. From the equation, it is obvious that any one of the variables can be determined if the other two variables are known. As electrons flow through ordinary conductors, they encounter forces that oppose their motion. This opposition to the electron flow is called resistance. The force that moves the electrons is the voltage. And the quantity of electrons that move is the current. Every electronics company must use Ohm's law to design its products. Electric power companies use this law when they design high-voltage power lines. A wire must be able to carry enough current at a given voltage, and Ohm's law is used to determine how much resistance that wire can have.

Misconception: Students may think that Ohm's law applies only to direct-current circuits. In fact, it applies to any circuit in which current is flowing through a resistor, even alternating current.

Process Skills

measuring in SI, using numbers, collecting and organizing data, interpreting data, making and using graphs, analyzing

Preparation Hints

- As an alternative to using the CBL unit, a 0–5 VDC voltmeter and a 0–50 mA ammeter can be substituted in place of the current and voltage probes. An excellent source for an inexpensive, high-quality voltmeter is Radio Shack. The versatile multitester unit provides several voltage ranges and current ranges from the same meter. The multitester can measure resistance if it is powered by a battery, which, of course, you would not include if students are expected to find the values of unknown resistors. This multitester has a fairly low resistance. When used to measure voltage, it will introduce an error of 1–2% for the resistor values used in the experiment.

Materials

power supply, 0–5 VDC or two 1.5-V batteries
knife switch
connecting wire
CBL unit
graphing calculator with PHYSICS CBL software program
two CBL-DIN adapters
voltage and current probe with dual-channel amplifier box
link cable
resistors: 100-Ω, 150-Ω, and 220-Ω, with 0.5-W or 1.0-W rating

Procedure Hints

- Variable power supplies work well for this experiment. Remind students not to exceed 5 V potential difference.

- The recommended resistors are readily available and inexpensive, particularly at an electrical supply house or Radio Shack. Use of milliammeters and these resistors reduces the circuit currents; however, the resistors can still become warm. Caution students not to touch resistors when current flows in the circuit.

- Create unknowns by covering known resistors with heat-shrink tubing, so the color bands are not visible. To prevent small current readings, which are difficult to measure, try not to exceed values of about 470Ω for unknowns. A couple of resistors can be placed in parallel or in series to create various resistance values.

Data and Observations

Resistor	Printed value of resistor (Ω)	Tolerance range (+/– %)	Voltage (V)	Current (mA)	Current (A)	Resistance (Ω)
		Table 1 (Sample data are given.)				
R_1	100	20	1.5	14	0.014	110
	100	20	5.0	49	0.049	100
R_2	150	20	1.5	10	0.010	150
	150	20	5.0	30	0.030	170
R_3	220	20	1.5	7	0.007	210
	220	20	5.0	24	0.024	210

Analysis and Conclusions

1. See Resistance column of the table.

2. R_1 error: $(105 - 100)(100\%)/100 = 5\%$
   R_2 error: $(160 - 150)(100\%)/150 = 7\%$
   R_3 error: $(220 - 210)(100\%)/220 = 5\%$

   These sample values agree fairly well; however, student values may be more divergent.

3. The resistors may have overheated, causing the values to change. The voltmeter may introduce some error in the circuit, since it requires some current to deflect the needle.

4. The ammeter must be in series with the circuit element.

5. The voltmeter must be in parallel with the circuit element to be measured.

6. The current in a circuit is equal to the voltage divided by the resistance.

Extension and Application

1. Answers will vary. Graphs of voltage versus current should be linear and should have a positive slope.

2. $$R = \frac{V}{I} = \frac{120\ V}{0.5\ A} = 240\ \Omega$$

   $$I = \frac{V}{R} = \frac{120\ V}{21\Omega} = 5.7\ A$$

22-2

Electric Equivalent of Heat

Introduction

Energy can not be destroyed; it can only change form. Thermal energy is always produced when an electric current flows through a resistance. For example, semiconductor devices—such as those in computers—and electric motors generate heat. Other types of energy may also be produced by an electric current. An incandescent lightbulb is an example of a resistance that changes electric energy into thermal energy and light energy. If the amount of electric energy passing through a conductor is relatively small, then only heat energy will be produced. Most electric and electronic devices operate poorly when they are too hot, so it is important to know how much heat is generated from electric current. In this experiment, an electric current passing through a resistance heats a known mass of water. This quantity of thermal energy is compared with the electric energy that was used to heat the water.

Process Skills

measuring in SI, using numbers, comparing and contrasting

Preparation Hints

- A rheostat and a constant-voltage power supply may be substituted for the variable power supply.

Materials

polystyrene cup
power supply, 0–15 VDC
connecting wire
ammeter, 0–5 A
rheostat
voltmeter, 0–15 VDC
thermometer
balance
stopwatch or watch with second hand

Procedure Hints

- Remind students to stir the water gently to prevent localized heating.

- A large rheostat may be used to control the current flow from a constant-voltage power supply. Most new power supplies are variable and have a built-in rheostat.

Data and Observations

Table 1 (Sample data are given.)	
Mass of polystyrene cup	1.1 g
Mass of water and cup	189.3 g
Mass of cold water	188.2 g
Initial temperature of water	14.1°C
Room temperature	25.0°C
Final temperature of water	36.2°C
Change in temperature of water	22.1°C
Average current	2.5 A
Average voltage	10.0 V

Table 2 (Sample data are given.)		
Time (min)	Current (A)	Voltage (V)
1	2.5	10.0
2	2.5	10.1
3	2.5	10.1
4	2.5	10.0
5	2.5	10.0
6	2.5	10.1
7	2.5	10.0
8	2.6	10.3
9	2.4	9.7
10	2.5	10.0
11	2.5	10.1
12	2.5	10.0
13		
14		
15		

Analysis and Conclusions

1. $E = IVt = (2.5 \text{ A})(10.0 \text{ V})(720 \text{ s}) = 18\ 000 \text{ J}$

2. $Q_w = m_w C_w \Delta T_w =$
   $(188.2 \text{ g})(4.18 \text{ J/g·°C})(22.1°C) = 17\ 400 \text{ J}$

3. % difference $= \dfrac{(E - Q_w)(100\%)}{E}$

   $\dfrac{(18\ 000 - 17\ 400)(100\%)}{18\ 000} = 3.3\%$

4. Not accounted for were the energy used to raise the temperature of the coil and associated metal supports and the energy used to raise the temperature of the thermometer.

5. Yes. Students should realize that the polystyrene cup calorimeter is not an ideal setup.

6. % conversion $= (17\ 400 \text{ J}/18\ 000 \text{ J})(100\%)$ $= 97\%$. Answers will vary, but should be no less than 90%.

Extension and Application

1. $\Delta U = mgh = (1 \text{ kg})(9.8 \text{ m/s}^2)(50 \text{ m}) =$
   $490 \text{ J} = \Delta \text{thermal energy} (\Delta Q)$
   $\Delta T = \Delta Q/mC = 490 \text{ J}/(1000\text{g})(4.18 \text{ J/g·°C})$
   $= 0.12°C$
   This is not sufficient to produce a warm shower.

2. $Q_w = (4.18 \text{ J/g·°C})(180 \text{ g})(89°C) = 67\ 000 \text{ J}$
   $t = 67\ 000 \text{ J}/200 \text{ J/s} = 335 \text{ s or 5 min 35 s}$

3. $100\% - 16\% = 84\%$ converted to heat, so $(0.84)(100 \text{ W}) = 84 \text{ W}$ expended as heat. Since there is 1 J/s per watt, $(84 \text{ W})(1 \text{ J/s/W}) = 84 \text{ J/s}$.

23-1

Series Resistance

Introduction

Students will design a circuit that has several resistors in series. They will measure the voltage across each resistor in the circuit, and perhaps the current flowing through each resistor. They will measure the voltage across the entire circuit, as well as the total current in the circuit. Students will learn that the sum of the voltages appearing across the individual resistors equals the voltage across the entire circuit and that the current in any one resistor is the same as the current in every other resistor in a series circuit. In this lab, students will apply the relationship $R = V/I$, discussed in Chapter 23.

Process Skills

hypothesizing; making and using tables; collecting and organizing data; interpreting data; analyzing

Possible Hypotheses

The correct hypothesis is that a portion of the total voltage crosses each resistor, and the sum of those voltages is the same as the voltage across all of the resistors. The current through each resistor in a series circuit is the same.

Possible Materials

CBL unit
link cable
voltage and current probe with dual-channel amplifier box
2 CBL-DIN adapters
graphing calculator with PHYSICS CBL software program
power supply with variable voltage
wires with clips
resistors, 100 Ω, 150 Ω, 220 Ω, 470 Ω
voltmeter, 0–5 VDC or 0–15 VDC
ammeter, 0–50 mA or 0–100 mA
knife switch

Possible Procedures

The following procedure works well: Assemble a circuit consisting of three resistors in series with a power supply. Place an ammeter or current probe in the circuit to measure the total current in the circuit. With a voltmeter or voltage probe, measure the voltage across each individual resistor and the total voltage across the circuit. Use $R = V/I$ to calculate the total circuit resistance and the actual value of each resistor. Calculations and measurements should agree.

Teaching Suggestions

- Warn students that, if they apply too much voltage, the resistors will get hot.

- Remind students to check the polarity of their meters before applying power to their circuits.

- Remind students to convert milliamperes to amperes before doing their calculations.

- Students should measure the total voltage across only the resistors. If they include the ammeter, their results will have significant errors.

- When using the CBL unit, do not exceed 5 volts with the dual-channel amplifier.

- For CBL use, as a hint, refer students to the previous lab (22-1) for instructions on how to set up and use the dual-channel amplifier with voltage and current probes.

Data and Observations

3. The total voltage drop would stay the same, but the individual resistors would each have a greater drop.

4. The total current in a series circuit is directly proportional to the voltage applied to the entire circuit, and is inversely proportional to the total resistance of the circuit.

5. The equivalent resistance of a series circuit is equal to the sum of the individual resistances.

Apply

1. Resistance of the set of bulbs:

$$R = \frac{V}{I}$$

$$= \frac{120V}{0.5A} = 240 \ \Omega$$

Resistance of one bulb:

$$\frac{R}{100} = \frac{240 \ \Omega}{100} = 24 \ \Omega$$

The voltage across each bulb is the applied voltage (120 V) divided by 100, or 1.2 V.

2. Answers may vary. One method is to wire the unmarked resistor in series with a resistor of known value and the battery. Measure the voltage across each resistor and the battery. Use $R = V/I$ to determine the value of the unmarked resistor.

Table 1 (Sample data are given.)							
Resistance Value and Voltage							
Resistance value (Ω)			Total current	Total voltage	Voltage (V)		
Resistor 1	Resistor 2	Resistor 3	(mA)	(V)	V1	V2	V3
100	220	150	13.0	6.0	1.3	2.8	1.9

Analyze and Conclude

1. The sum of the individual voltage drops equals the total voltage across the series circuit.

2. Yes, the voltage drop is proportional to the resistance. A larger resistance has a larger voltage drop across it.

23-2

Parallel Resistance

Introduction

Each resistor in a parallel resistance circuit always has the same potential difference, and each resistor provides a path for current. The relationship, $R = V/I$ can be used to calculate the total resistance of the parallel circuit using the sum of the applied voltage and the current flowing in the resistors. The result is the effective resistance of the parallel circuit. The total resistance of resistors in parallel is always less than the lowest value of an individual resistor. Why use several resistors in parallel when they could be replaced by a single resistor? Resistors come in standard values, while a circuit, particularly an electronic one, may need a nonstandard value. If a circuit needs to be adjusted, it is usually less expensive to use several resistors in parallel than to use an adjustable resistor.

Process Skills

observing and inferring, measuring in SI, using numbers, comparing and contrasting, collecting and organizing data, interpreting data

Materials

DC power supply or dry cells
3 resistors, 0.5-W, 150–330-Ω range, such as 180 Ω, 220 Ω, and 330 Ω
connecting wire
knife switch
voltmeter, 0–5 VDC
ammeter, 0–50 mA or 0–100 mA

Procedure Hints

- Remind students to convert milliamps to amps before beginning their calculations.

- Caution students to use care when touching the resistors and to leave switches closed for no more than a few seconds at a time.

- Depending on the internal resistance of the ammeters, data may not match perfectly.

- Allow sufficient time for this experiment. It takes time to move the meters about and obtain circuit measurements.

Data and Observations

Table 1 (Sample data are given.)

R_1 (Ω)	Ammeter reading (mA)	Voltmeter reading (V)
180	173	3.0
Tolerance (%) 10		

Table 2 (Sample data are given.)

R_1 (Ω)	R_2 (Ω)	Ammeter reading (mA)			Voltmeter reading (V)		
		I	I_1	I_2	V	V_1	V_2
180	220	30	17	14	3	3	3
Tolerance (%)							
10	10						

Table 3 (Sample data are given.)

R_1 (Ω)	R_2 (Ω)	R_3 (Ω)	Ammeter reading (mA)				Voltmeter reading (V)			
			I	I_1	I_2	I_3	V	V_1	V_2	V_3
180	220	330	39	17	14	10	3	3	3	3
Tolerance (%)										
10	10	10								

Analysis and Conclusions

1. $R_1 = \dfrac{3.0 \text{ V}}{0.017 \text{ A}} = 180 \ \Omega$

   Answers will vary; the sample data yield a result within the tolerance range.

2. a. $R = \dfrac{V}{I} = \dfrac{3.0 \text{ V}}{0.03 \text{ A}} = 100 \ \Omega$

   b. $I = 0.017 \text{ A} + 0.014 \text{ A} = 0.031 \text{ A}$

   c. $R_1 = \dfrac{3.0 \text{ V}}{0.017 \text{ A}} = 180 \ \Omega$

   d. $R_2 = \dfrac{3.0 \text{ V}}{0.014 \text{ A}} = 210 \ \Omega$

   e. $R = 97 \ \Omega$

3. Student values may have a greater difference, depending upon the quality of the ammeters used.

   a. The measured current should equal the sum of the currents (or very nearly).

   b. Using the sample readings, the calculated equivalent resistance (97 Ω) equals the measured equivalent resistance (100 Ω) within the tolerance of the resistors.

4. a. $R = \dfrac{3.0 \text{ V}}{0.039 \text{ A}} = 77 \ \Omega$

b. $I = 0.017 \text{ A} + 0.014 \text{ A} + 0.010 \text{ A}$

   $= 0.041 \text{ A}$

c. $R_1 = \dfrac{3.0 \text{ V}}{0.017 \text{ A}} = 180 \text{ } \Omega$

d. $R_2 = \dfrac{3.0 \text{ V}}{0.014 \text{ A}} = 210 \text{ } \Omega$

e. $R_3 = \dfrac{3.0 \text{ V}}{0.010 \text{ A}} = 300 \text{ } \Omega$

f. $R = 73 \text{ } \Omega$

5. Student values may have a greater difference, depending on the quality of the ammeters used.

   a. The two currents are very nearly equal. (They should be equal.)

   b. Using the sample readings, the calculated equivalent resistance (73 Ω) equals the measured equivalent resistance (77 Ω) within the tolerance of the resistors.

6. The sum of the current in the branches of a parallel circuit equals the total current in the circuit.

7. The voltage drop across each branch of a parallel circuit is the same as the voltage drop across the entire circuit.

8. The total current increases.

Extension and Application

1. The shunt resistance is

$$\frac{1}{R} = \frac{1}{R_1} + \frac{1}{R_2} = \frac{1}{0.5000 \text{ } \Omega}$$

$$= \frac{1}{500.0 \text{ } \Omega} + \frac{1}{R_s}.$$

$$\frac{1}{R_s} = 2.000 \text{ } \Omega - 0.0020 \text{ } \Omega = 1.998 \text{ } \Omega$$

$$R_s = \frac{1}{1.998} = 0.5005 \text{ } \Omega$$

2. The 10 000-Ω voltmeter would not be a good choice because the effective circuit resistance is 909.1 Ω and the voltage measured would be low by over 9.0%. The 1 000 000-Ω voltmeter would be a good choice because the effective circuit resistance is 999.0 Ω and the voltage measured

would be low by about 0.1%.

3. Answers will vary. Sample answer: Reduce the circuit to two parallel sets, each having four 100-Ω resistors in parallel connected in series with a pair of 100-Ω resistors in parallel. This reduces to two 75-Ω resistors in parallel, which is equivalent to one 37.5-Ω resistor in series with the 1.0-V battery.

The current $= I = \dfrac{V}{R} = \dfrac{1.0 \text{ V}}{37.5 \text{ } \Omega} = 27 \text{ mA}.$

24–1

The Nature of Magnetism

Introduction

Because the north-pointing end of a compass needle has a north magnetic polarity, you can use a compass to find the south pole of a magnet. The magnetic field flows out of the north pole of the magnet, around the outside of the magnet, and into its south pole. The magnetic field lines are continuous outside and inside the magnet. When one end of a metal object, such as an iron nail, is attached to one pole of a magnet, the induced magnetism generates a magnetic pole at the free end of the object that is of the same polarity as the end of the magnet that is attached to the object.

In this lab, students will apply concepts of magnetism and polarity covered in Chapter 24 of their textbook.

Process Skills

hypothesizing; observing and inferring; interpreting scientific illustrations

Possible Materials

2 bar magnets
sheet of paper or thin cardboard
magnetic compass
iron filings
iron nail

Possible Hypotheses

The hypothesis that should be supported by the results is that the compass will point to the south pole of the magnet and that a map of the magnetic field lines will indicate the attraction. Other hypotheses are that compasses point to the north pole of a magnet or that they aren't affected by ordinary magnets. These hypotheses should not be supported by the results.

Possible Procedures

The following procedure works well: Hold one end of the magnet close to a compass and observe which way the compass points. To observe the magnetic field pattern of a magnet, lay a sheet of paper on the magnet and sprinkle a thin layer of iron filings on the paper. The filings will align their long axes with the magnetic field of the magnet. Record the field lines in a drawing. The same method can be used to show the field between two magnets. To trace the magnetic field around a magnet, move a compass around the magnet and record the direction of the north pole of the compass at different places.

Teaching Suggestions

- Make sure that the magnets are in good condition so that they will make distinct field patterns.

- To prevent the iron filings from becoming attached to the magnets, put the magnets in a plastic bag.

- If students aren't getting distinct field patterns, they may be using too many iron filings. Remind them to tap the paper lightly to get the filings to align with the field.

Data and Observations

Magnetic field lines between poles

Direction and shape of magnetic field lines

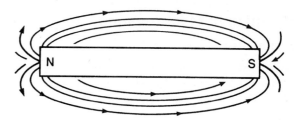

Analyze and Conclude

1. The magnetic field is most concentrated near the poles. The lines between like poles bend away from each other. The lines between unlike poles join with each other to form continuous lines.

2. The compass points toward the south pole of the bar magnet.

3. The magnetic field lines between the compass and the south pole of the magnet show that they are unlike poles, which attract each other.

4. The polarity of the induced magnetism at the free end of the nail is the same as the end of the magnet that is attached to the nail.

Apply

1.

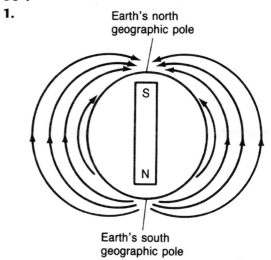

Earth's north geographic pole

Earth's south geographic pole

2. Answers will vary. Sample answer: Use a compass to see what end the magnet attracts. Put the magnet among iron filings to see field lines. Most refrigerator magnets will show a pattern of alternating bands. The spacing of the bands varies considerably; some are so close that the magnet may seem to have a single pole.

24-2

Principles of Electromagnetism

Introduction

Direct current flowing through a wire generates a magnetic field around the wire. The intensity of the magnetic field is directly proportional to the amount of current flowing through the wire. If the wire forms a coil, the magnetic field increases in direct proportion to the number of turns in the coil. A coil with an iron core can be a hundred to several thousand times what the field would be in an air-core coil carrying the same current. Some small electromagnets are capable of lifting a mass of 50 kg or more with a current of only 0.001 A.

Process Skills

observing and inferring, formulating models, interpreting scientific illustrations

Preparation Hints

- If students use batteries, they need a rheostat to adjust the current. The batteries should be fresh because the experiment requires high current.

- A good substitute for an iron nail is a large-diameter steel bolt.

Materials

small magnetic compass
DC power supply, 0–5 A (or 3 dry cells)
ammeter, 0–5 A
14–18 gauge wire, 1.5 m
large iron nail
knife switch
ring stand
ring-stand clamp
iron filings
cardboard sheet, $8\frac{1}{2} \times 11$ inches
plain white paper
meterstick
masking tape
box of steel paper clips

Procedure Hints

- Review the right-hand rule with students before the lab.

- Remind students to leave the switches closed only long enough to make observations.

- If students have trouble with current fluctu-ations, they should check that the wire connections are tight. Other causes might be a bad power supply, a dirty rheostat, or weak batteries.

Data and Observations

A. The Field Around a Long, Straight Wire

With current flowing upward, the north pole points counterclockwise about the wire.

With current flowing downward, the north pole points clockwise about the wire.

B. Strength of the Field

The filings form circular patterns around the wire. The smaller the current, the less distinct the patterns. With 0.5 A of current, almost no pattern is observed.

C. The Field Around a Coil

The compass needle lines up as indicated by the labels in the figure in Analysis and Conclusions below.

D. An Electromagnet

1. Sample electromagnet picks up six clips.

2. Sample electromagnet picks up ten clips; increasing the number of loops increases the magnet's strength.

3. Sample electromagnet picks up 16 clips; increasing the current increases the magnet's strength.

Analysis and Conclusions

1. The direction of the magnetic field follows the direction of the fingers of the right hand when the right thumb points in the direction of the conventional current (from positive to negative).

2. Increasing current increases electromagnetic strength.

4. The number of turns, the current strength, and the type of core determine the strength of an electromagnet.

5. A bar magnet maintains its magnetism for a long time, while an electromagnet loses its magnetism as soon as the current flow stops.

Extension and Application

1. Continuously applied current: strong electromagnets—such as in junkyards, magnetic-resonance-imaging equipment

Intermittently applied current: telephone ringers, doorbells, buzzers, door locks, electric switches (relays)

24-3

Variation in the Strength of Electromagnets

Introduction

Students will be making electromagnets and using them to transfer as many BBs as possible from one cup to another. Because each BB has about the same mass and the force of gravity, which opposes the force from the magnetic field, is constant, the number of BBs picked up is proportional to the strength of the magnetic field of the electromagnet.

Inaccuracies arise from the inability to hold fractions of a BB and the magnetic properties of the bolts used for the core of the electromagnets. Most bolts are not iron but an iron alloy, and some perform better than others. The strength of the electromagnets also varies depending on how the coil is wound and where it is on the bolt. A coil with turns close together produces a stronger field than a coil with space between the turns. A coil on the bolt close to the end used to pick up the BBs has a stronger field than a coil at the opposite end or in the middle of the bolt. This dependence on position results partly from the rather poor magnetic properties of the bolt material.

Process Skills

hypothesizing, communicating, experimenting, analyzing data, using numbers, making and using tables, defining operationally, formulating models

Possible Hypotheses

The correct hypothesis is that the strength of the magnetic field of an electromagnet is directly proportional to the number of wire turns in the coil of the electromagnet.

Possible Procedures

One procedure that works well is to make several electromagnets by winding wire around identical iron bolts. Use a different number of turns for each magnet and label each one. Connect an electromagnet to a source of current and use the electromagnet to pick up BBs from one cup. Transfer the BBs to another cup, count the BBs, and record the count. Repeat the procedure for all the electromagnets, using the same amount of current each time.

Possible Materials

insulated magnet wire
iron bolts
paper cups
ammeter, 0–5 A
DC power supply, 0–5 A
iron BBs
masking tape
connecting wire
marking pen

Teaching Suggestions

- Tell students to pick up any BBs that drop on the floor so that they do not slip on them.

- Caution students that the wire in their electromagnets can get hot if they use too much current.

- Advise students to insulate their bolts with masking tape to prevent shorting out the coil, particularly where it passes over the threads of the bolt.

- Students should wind the wire tightly and evenly around the bolts.

Data and Observations

Table 1 (Sample data are given.)			
Electro-magnet	Wire turns	BBs picked up	Value of k
A	10	4	2.5
B	20	9	2.2
C	30	13	2.3
D	40	19	2.1
E	50	23	2.2

Analyze and Conclude

1. The strength of the magnetic field increases as the number of wire turns increases.

2. $y = kx$ or $x = y/k$

3. The calculated constants will vary, but each set of electromagnets should have about the same value. The average value of k for the sample data is approximately 2.3.

4. Answers will vary, but should indicate that the current in the coil should be the same for all trials because the strength of the magnetic field is proportional to the current.

5. ~ 3500 coils

Apply

1. Answers will vary. Sample answer: The electromagnets have many wire turns, and the core material is a very good magnetic material. If the electromagnet has a U-shaped core, both its poles can exert force on the door.

25-1

Electromagnetic Induction 1

Introduction

Electromagnetic induction is a process in which electric current is generated in a conductor by a moving magnetic field. This process is the principle behind electric generators. Electric current is generated only when relative motion exists between the conductor and the magnetic field. The current generated varies, depending on the strength of the magnetic field, the number of con-

ductors, and the rate at which the magnetic field moves relative to the conductor. In other words, a more powerful magnetic field generates more current in the same conductor, and more conductors, in the form of a coil, generate more current from a given magnetic field. In this experiment, students investigate the generation of electrical current when a magnetic field cuts through a conductor.

Process Skills

observing and inferring, comparing and contrasting, communicating, interpreting data

Preparation Hints

- Students can wind the coils around cardboard tubes, such as toilet-paper tubes or paper-towel tubes, using 20- or 22-gauge insulated copper wire. A few pieces of masking or electric tape keep the coils together. Commercially made coils are available, or use a galvanoscope block with wound coils.

Materials

galvanometer with centered zero
1-turn coil of wire
25-turn coil of wire
100-turn coil of wire
2 bar magnets
connecting wire

Procedure Hints

- Help students understand that, whereas the conversion of electric energy to mechanical energy in a magnetic field is the principle of the electric motor, the conversion of mechanical energy to electric energy in a magnetic field is the principle of the electric generator. Emphasize that this correspondence of opposites prevails in many areas of science and provides guidance and structure to researchers and students alike.

- Remind students to keep magnets away from their wristwatches.

Data and Observations

Table 1 (Sample data are given.)	
Observations of needle directions	
Magnet into 1-turn coil	The galvanometer barely moves. Answers will vary, depending on the strength of the magnet and the sensitivity of the galvanometer.
Magnet into 25-turn coil	The galvanometer needle should move, indicating that a current was produced.
Magnet into 100-turn coil	A much stronger current is produced, compared to that produced by the 25-turn coil.
Magnet into and out of coil	The galvanometer needle deflects in one direction when the magnet is thrust into the coil and in the opposite direction when the magnet is pulled out. The needle deflects in opposite directions when the north or south pole is thrust into the coil, due to the different directions of the magnetic fields.
Two magnets into 100-turn coil, differnt velocities	The stronger magnet produces a stronger current. The greater the speed of the magnets' motion, the stronger the current.
Magnets in 100-turn coil: stationary and moving coil	No current is induced when the magnets are stationary. A current is produced when the coil moves over the magnets.

Analysis and Conclusions

1. The end of the coil into which the north pole of the magnet is thrust becomes a south pole, and then reverses when the magnet is pulled out. An induced current acts in such a direction that its magnetic properties oppose the change that induces the current.

2. The *EMF,* measured in volts, depends on the strength of the magnetic field, the number of loops in the coil, and the speed at which the wire cuts through the magnetic flux. If the moving wire is part of a closed circuit, the induced current is equal to the voltage divided by the resistance of the circuit.

3. From Ohm's law, $V = IR$. If resistance is constant, the voltage (*EMF*) is proportional to the current. This experiment showed that the strength of the magnetic field, B, affected the magnitude of the current; the number of turns, L, affected the current; and the rate of movement, v, of the wire through the magnetic field affected the current. Therefore, $EMF \propto B$, $EMF \propto L$, and $EMF \propto v$, and the equation, $EMF = BLv$, is substantiated.

4. No current flows in the wire. An electric current is produced only when the wire cuts the magnetic field lines.

5. No matter what its source, each field exerts a force, $q'E$, on a test charge, q'.

Extension and Application

1. $EMF = BLv \sin \theta$
   $= (0.75 \times 10^{-2} \text{ T})(0.40 \text{ m})(5.0 \text{ m/s})(\sin 45°)$
   $= 1.1 \times 10^{-2} \text{ V}$

2. *EMF* is greatest when $\theta = 90°$ and diminishes to zero as θ approaches $0°$.

3. The output is an alternating current that varies sinusoidally over a period of time; the amplitude of the current changes with the sine of the angle between v and B. In the United States, the frequency of the alternating current is 60 cycles/s; the current changes direction 120 times/s.

4. A changing magnetic field appears in the iron core due to the alternating current in the primary coil. Through the iron core, the field moves over the loops of the secondary coil, causing an induced current, and an induced *EMF,* to appear in that coil. The transformer does not work with direct current since no changing magnetic field is produced; thus, no current can be induced.

Electromagnetic Induction 2

Introduction

Electrical current is generated when a wire moves through a magnetic field or when a magnetic field cuts through a wire. Mechanical motion is generated when an electrical current flows through a wire in a magnetic field. A generator usually has a coil of wire that is rotated in a magnetic field. A motor usually has electric current flowing through a coil of wire that is free to move in a magnetic field. Some newer types of generators and motors have the coils stationary while the magnets move. An example of this type of construction is the automobile alternator. The moving coil of wire in a motor also acts as a generator that attempts to supply current back to the source. This phenomenon is defined by Lenz's law. This back-*EMF* opposes the source and reduces the total current flowing in the motor coils. In this experiment, a simple generator and a simple motor are constructed, examined, and compared.

Process Skills

observing and inferring, comparing and contrasting, communicating

Preparation Hints

- If the horseshoe magnets are weak, try remagnetizing them. To increase the magnetic field strength, two magnets may be placed side by side, with like poles together, and secured with pieces of tape.

- Different lab groups could wind different numbers of turns to determine which size coil works best.

- Smaller wire, such as 26- or 28-gauge wire, makes the best coils because of its lighter weight.

- It is useful to discuss Lenz's law before the experiment.

Materials

100-mL beaker or cardboard tube to form coil
22-, 24-, 26-, or 28-gauge enameled magnet wire, 60 m
galvanometer with centered zero (or 500-μA to 0 to 500-μA meter)
ring stand
2 ring-stand clamps
masking or electric tape
2 horseshoe magnets
fine sandpaper, 2 cm × 2 cm
connecting wire

Data and Observations

1. The swinging coil has an induced current that changes direction as the swinging coil changes direction. The coil generates 200–500 μA per swing. The faster the coil swings, the stronger the induced current is. The stronger the magnetic field, the stronger the induced current is.

2. When one coil starts swinging, the other also starts to swing. The coils swing in opposite directions and gradually come to rest.

3. The unhooked coil swings faster than it does when attached to the other coil.

4. The galvanometer shows that current flows from one coil through the galvanometer to the other coil, causing it to move. When the second coil swings back, it generates current in the opposite direction; this current flows back to the first coil. The entire process continues back and forth until resistance brings the system to a rest.

Analysis and Conclusions

1. Factors affecting the strength of the induced current include the relative speed of the wire through the magnetic field and the strength of the magnetic field.

2. The current produced when one coil moves flows to the other coil through the connecting wires. This current interacts with the magnetic field surrounding the second coil, causing it to move. When the second coil swings back, it generates current in the opposite direction, which flows back to the first coil. The entire process continues until resistance brings the system to rest.

3. The unattached coil swings faster. Lenz's law states that the magnetic effects produced by the induced current oppose the change in flux that caused the current. When the generator coil is attached to the second coil, a load exists on the generator, which creates an opposing force.

4. The first coil set into motion acts like the generator; its induced current causes the second coil, which is the motor, to swing. The process reverses itself as the coils swing in the opposite direction.

Extension and Application

1. When the lights are turned on, a larger current flows from the generator. When the current increases, the opposing force also increases, and it is more difficult to turn the generator. A larger opposing force means that a larger amount of mechanical energy is needed to turn the generator.

26-1

Mass of an Electron

Introduction

The force exerted on an electron moving perpendicularly to a uniform magnetic field is defined by Newton's second law of motion. All that is necessary to determine the mass-to-charge ratio is the velocity of the electron, the radius of the curved path taken by the electron, and the strength of the magnetic field that causes the curvature. Then if the charge is known, the mass can be calculated. When Thomson performed his experiments, the charge on an electron was unknown, so all he could determine was the ratio of mass to charge. Years later the charge on an electron was determined, and with that knowledge came the mass of the electron. In this experiment, students will use a procedure like Thomson's to determine the mass-to-charge ratio for an electron. From that ratio and the known charge on an electron, they will calculate the mass of an electron.

Process Skills

observing and inferring, measuring in SI, using numbers, interpreting data, comparing and contrasting

Preparation Hints

- The 6E5-style tube is available from major electronics supply houses and Radio Shack. Usually this item is a special order.

- A suitable coil may be constructed by winding 180–200 turns of 16–20-gauge enameled magnet wire in two layers on a form about 6.5 cm in diameter. The winding will be 8–10 cm long.

Materials

CBL unit
link cable
graphing calculator with PHYSICS CBL software program
magnetic field sensor
CBL-DIN adapter
mass-of-the-electron apparatus
   (6E5 style tuning-eye tube)
DC power source, 0–5 A
   (or 6-VDC power supply and rheostat)
DC power source, 90–250 V
DC or AC filament power supply, 6 V
air-core solenoid
connecting wire
ammeter, 0–5 A
3 dowel rods, different diameters
voltmeter, 0–250 VDC

Procedure Hints

- Caution students not to use any metal object as a substitute for the dowel. A metal object can seriously alter the deflection pattern of the tube.

- Students should work quickly while current is applied to the coil; the warming of the wire may decrease the current.

- Instead of using the CBL unit to measure the magnetic field strength, an approximation of the field strength, can be made using the following relationship:
$B = I \times t \times l \times 2.52 \times 10^{-4}$ where the magnetic field induction, B, is in Teslas (N/A·m) ; the current in the coil, I, is in amperes, the number of turns of wire per centimeter of the coil is t, and l is the number of layers of wire on the coil.

Data and Observations

Trial	Coil current, I (A)	Magnetic induction, B (T, N/A·m)	Potential difference, V (V)	Radius, r (m)
1	1.4	0.0062	160	0.0075
2				
3				

Table 1 (Sample data are given.)

Trial	B^2 (N/A·m)^2	r^2 (m^2)	Charge on one electron, q (C)	mass/charge (kg/C)	Mass of electron (kg)
1	3.8×10^{-5}	5.63×10^{-5}	1.6×10^{-19}	6.69×10^{-12}	1.1×10^{-30}
2					
3					

Table 2 (Sample data are given.)

Analysis and Conclusions

1. See Table 2.

2. 1.1×10^{-30} kg for the sample data.

3. Relative error = $(1.1 \times 10^{-30}$ kg $- 9.1 \times 10^{-31}$ kg$)$ $(100\%)/(9.1 \times 10^{-31}$ kg$)$ = 9%. Student error will vary depending on the accuracy of the coil data and the accuracy in approximating the radius. In general, all answers should be within an order of magnitude.

4. Regardless of even a large error, such as 50–100% or more, the mass of an electron has been approximated with relatively simple equipment, and it is very, very small.

Extension and Application

1. The magnetic induction of Earth is about 100 times smaller than that of the coil. To effect the same displacement, a very large tube would be needed because Earth's magnetic field would cause a very small deflection, which is probably not measurable in one of our laboratory tubes. Students might propose using a larger tube or using a long, straight tube and measuring the deflection parallel with the field to compare it to the deflection perpendicular to the field.

27-1

Plank's Constant

Introduction

A light-emitting diode (LED) is a semiconductor diode that generates light from electric energy. This type of diode turns Planck's assumptions about the energy of vibration of atoms in a solid into a practical device. An LED is constructed in a way similar to rectifier diodes. The important difference is that an LED generates light when there is a current through it, whereas a rectifier diode generates heat. LEDs emit light over a very narrow band of wavelengths, usually only about 20–30 nm. The wavelength of the emitted light is a function of the semiconductor composition and the applied voltage. In this experiment, measurements are taken of the voltage across and the current through light-emitting diodes of various colors. From the data, a value is calculated for Planck's constant for each color of LED. The experiment shows the relationship between the electric energy, measured in volts, put into the LED and the wavelength of the light given off by that LED.

Process Skills

collecting and organizing data, formulating models, interpreting data, making and using graphs, using numbers

Preparation Hints

- A variable-power supply can be used instead of the battery; however, it should provide well-filtered, clean DC power. Most new variable power supplies meet these criteria.

Materials

green, red, and yellow LEDs
2 1.5-V C-size batteries or 3-V power supply
battery holder
22-Ω resistor
1000-Ω potentiometer
connecting wire
ammeter, 0–50 mA DC
voltmeter, 0–5 VDC
knife switch

Procedure Hints

- The packing material of many LEDs displays the wavelength of their emitted radiation. If you have some LEDs that are not identified, however, the following table will help you approximate the wavelengths of light for typical LEDs.

Color	Wavelength (nm)
Red	635, 650, 660, 675, 690 (Standard red is 650 or 660.)
Yellow	590, 610
Green	550, 563, 570

- The 22-Ω resistor limits current to the LED, thus preventing instantaneous destruction of the LED if students accidentally apply too large a voltage. Resistance could be increased to as much as 33 Ω, if desired, or eliminated. If it is eliminated and a potential difference of 3 V is applied to the LED, excessive current will flow through the junction and destroy it.

Data and Observations

Table 1		
LED	LED color	Wavelength (nm)
1	red	650
2	yellow	590
3	green	563

Table 2 (Sample data are given.)					
LED1		**LED2**		**LED3**	
Voltage (V)	Current (mA)	Voltage (V)	Current (mA)	Voltage (V)	Current (mA)
1.50	0	1.50	0	1.50	0
1.55	0	1.55	0	1.55	0
1.60	0	1.60	0	1.60	0
1.65	1	1.65	0	1.65	0
1.70	2	1.70	0	1.70	0
1.75	3.7	1.75	0	1.75	0
1.80	5.0	1.80	1	1.80	0
1.85	7.5	1.85	1.7	1.85	1
1.90	11.0	1.90	3.1	1.90	1.8
1.95	13.5	1.95	5.5	1.95	3.2
2.00	17.5	2.00	7.0	2.00	5.1
2.05	22.0	2.05	13.5	2.05	9.0
2.10	25.0	2.10	17.0	2.10	13.0
2.15		2.15	21.0	2.15	15.0
2.20		2.20		2.20	18.0
2.25		2.25		2.25	21.0
2.30		2.30		2.30	23.0
2.35		2.35		2.35	

LED Current vs Voltage

This graph was produced with the Lab Partner graphing program.

Analysis and Conclusions

1. red voltage: 1.75 V
   yellow voltage: 1.95 V
   green voltage: 2.05 V

2. red LED, $h = qV\lambda/c = (1.6 \times 10^{-19}$ C$)$
   $(1.75$ V$)(6.50 \times 10^{-7}$ m$)/(3.0 \times 10^8$ m/s$)$
   $= 6.1 \times 10^{-34}$ J·s
   yellow LED, $h = qV\lambda/c = (1.6 \times 10^{-19}$ C$)$
   $(1.95$ V$)(5.90 \times 10^{-7}$ m$)/(3.0 \times 10^8$ m/s$)$
   $= 6.1 \times 10^{-34}$ J·s
   green LED, $h = qV\lambda/c = (1.6 \times 10^{-19}$ C$)$
   $(2.05$ V$)(5.63 \times 10^{-7}$ m$)/(3.0 \times 10^8$ m/s$)$
   $= 6.2 \times 10^{-34}$ J·s

3. red error $= (6.6 \times 10^{-34} - 6.1 \times 10^{-34})$
   $(100\%)/6.6 \times 10^{-34} = 8\%$
   yellow error $= 8\%$
   green error $= 6\%$
   Answers will vary but should be within
   10–15% of h.

4. A blue LED should have a voltage greater
   than a green LED, which may be 2.4–2.6 V.
   The infrared LED will have a voltage less
   than red LED, or less than 1.6 V, possibly
   1.2–1.4 V.

Extension and Application

1. It is not possible to generate white light
   from a single LED. To generate light with
   such a wide range of wavelengths would
   violate the wavelength-energy equation.
   Some students might propose that a device
   constructed from a red, a green, and a blue
   LED would give off light that appears white.

2. The LED is more durable because it has no
   fragile filament. Light from an incandescent
   lightbulb comprises the entire visible spectrum, whereas light from the LED has only
   one wavelength (monochromatic light). The
   wavelength of an LED is a very narrow band
   around its computed value, with a band-
   width of about 25–30 nm. When wavelength is closely controlled and the light is
   produced in a cavity at the junction, a laser
   diode results.

27-2

The Photoelectric Effect

Introduction

The photoelectric effect occurs when photons in
the form of light energy strike a metallic surface
and electrons are emitted. To produce a current
of photoelectrons, the frequency of the light
must be above a minimum value called the
threshold frequency. The emitted electrons have
an amount of energy proportional to the frequency of the light energy that generated them.
The energy of the emitted electrons can be measured as a potential difference or voltage. The
relationship between the frequency of the light,
thus its energy, shows that the higher the frequency, the greater the electric potential or voltage of the emitted electrons. From a graph of this
relationship, the threshold frequency for the
metal used as a photoelectric detector can be
determined. In this experiment, students shine
light of different frequencies on a photocell and
plot their results to determine the threshold frequency of the photocell.

Process Skills

formulating models, interpreting data, using
numbers, measuring in SI, making and interpreting data

Preparation Hints

- Check that your photoelectric-effect mod-
  ule(s) and associated equipment are in
  proper working order. Prepare a list of
  wavelengths of the filters that students will
  use in the experiment.

- The photoelectric-effect modules with associated amplifiers are available from most
  science suppliers or from Thornton Associ-
  ates. Different manufacturers have equipment that is usually compatible only with
  their own product line.

Materials

photoelectric-effect module with amplifier
voltmeter, 0–5 VDC
microammeter, 0–100 μA
colored filters: red, green, blue, yellow,
    and violet
40-W incandescent lightbulb in socket

Procedure Hints

- Because many filters are broad-band, allowing a range of frequencies to pass through, selection of a shorter wavelength works best.

- A sensitive voltmeter, such as a digital voltmeter, would more accurately measure the stopping voltage.

- Tell students that stray light entering the photocell from a source such as a window or room lights will cause errors in their measurements.

Data and Observations

Table 1 (Sample data are given.)

Filter color	Stopping voltage (V)	Energy (J)	Wavelength (m)	Frequency (Hz)
Red	0.20	3.2×10^{-20}	6.6×10^{-7}	4.5×10^{14}
Blue	0.45	7.2×10^{-20}	4.8×10^{-7}	6.3×10^{14}
Green	0.40	6.4×10^{-20}	5.2×10^{-7}	5.8×10^{14}
Yellow	0.41	6.6×10^{-20}	5.8×10^{-7}	5.2×10^{14}
Ultraviolet	1.45	2.3×10^{-19}	4.2×10^{-7}	7.2×10^{14}

Analysis and Conclusions

1. See Table 1.

2. Energy vs Frequency in the Photoelectric Effect

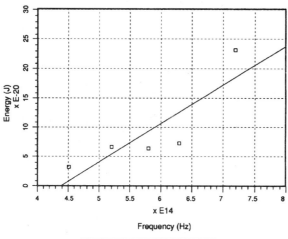

Linear Statistics: Energy (J) = f(Frequency (Hz))

Slope = 6.56338E-34   Correlation Coef. = 0.855863
Y Intercept = -2.87476E-19   X Intercept = 4.38E+14   Std. Error = 4.72638E-20

3. The x intersect represents the value of the threshold frequency at which photoelectrons are emitted with zero kinetic energy.

4. The slope of the graph should be Planck's constant. With the sample data, the slope equals 6.5×10^{-34}. This is only a 2 percent

difference from the accepted value of Planck's constant.

Extension and Application

1. From $E = hf = hc/\lambda$, you know that red light has a relatively low frequency (high wavelength) compared to other colors of light. Because red photons have less energy than green or blue photons, they are less likely to further expose the film.

28-1

Spectra

Introduction

Heating solid matter generates a continuous spectrum. Substances that are either vaporized in a flame or excited by electric discharge generate a discontinuous spectrum. This discontinuous spectrum consists of a series of bright lines when viewed with a spectroscope. Each element generates a unique set of these lines, and by comparing the set of lines from an unknown substance to those from known elements, you can identify the composition of the unknown substance.

Process Skills

observing and inferring, collecting and organizing data, comparing and contrasting, analyzing, defining operationally

Preparation Hints

- If you have several spectroscopes, groups of students could observe each spectrum tube simultaneously.

Materials

spectroscope

spectrum-tube power supply
spectrum tubes: argon, bromine, carbon dioxide, chlorine, helium, hydrogen, krypton, mercury vapor, neon, nitrogen, oxygen, and xenon
40-W incandescent lightbulb and socket
thermal mitt

Procedure Hints

- If two spectrum-tube power supplies are available, locate them in different parts of the room. Students can move from one to the other and collect data. You can change the spectrum tubes yourself, reducing the risk to students of shock or burns.

- The emission-spectra transparency from Chapter 28 (Transparency 50) can be used for a post lab discussion or for evaluation of data. Cover the element names to use the transparency as an evaluation instrument if you decide to give students unknowns for analysis and identification.

Data and Observations

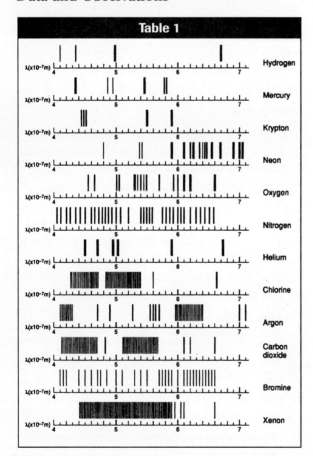

Table 1

Hydrogen

Mercury

Krypton

Neon

Oxygen

Nitrogen

Helium

Chlorine

Argon

Carbon dioxide

Bromine

Xenon

Analysis and Conclusions

1. The color of the glowing gas tube is different from that of the bright lines visible in the

spectroscope. The tube color is produced by a mixture of the individual spectral lines.

2. The spectral lines for each element vary in intensity. Some are much brighter than others.

3. The spectrum observed is identical to that of mercury gas. The fluorescent tube is filled with mercury gas.

Extension and Application

1. Xenon generates more light over most of the visible spectrum than other gases. This characteristic is most important for color photography.

2. Emission spectra of gases from the sun and other stars can be analyzed to determine the elements present in these distant objects.

3. Some practical uses might include analysis of unknown substances or determination of components of substances.

28-2

Measuring Electron Energy Changes

Introduction

When an electron absorbs heat, it moves to a higher-energy orbital. When the electron falls back to a more stable orbital, it releases the absorbed energy as light. Usually not all of the electrons in an atom gain sufficient heat energy from a flame to move to higher orbitals. In this experiment, students will use a diffraction grating to observe the spectral line of sodium atoms. They will use the Bragg equation, $\lambda = d \sin \theta$, to calculate the wavelength of the light emitted by sodium.

Process Skills

observing and inferring, using numbers, analyzing

Preparation Hints

- Have students work in groups of three.

- Supply the salt in small plastic bottles with lids.

- Dry powdered salt is necessary so that the dust from the salt is drawn up into the burner flame. If the salt is prepared several days before use, cover it tightly with plastic. A bag of silica gel will keep the salts dry.

- Baking soda (sodium hydrogen carbonate)

may be substituted for the sodium carbonate. It is less irritating than sodium carbonate and usually doesn't need drying.

Materials

apron
laboratory burner
lighter
2 metersticks
slide mount (optional)
masking tape
modeling clay
goggles
diffraction grating
cardboard with slit
white paper
sample of anhydrous sodium carbonate salts

Procedure Hints

- **CAUTION:** *Students should wear goggles and aprons for this laboratory. Students with long hair should tie their hair back. Sodium carbonate can irritate the skin. Use care when handling.*

- The number of lines per centimeter, n, is different for each type of diffraction grating. Write the correct value for n on the chalkboard for students to use in their calculations.

- Darken the room to prevent extraneous light from interfering with observations.

- Especially at the beginning of the experiment, make sure students have set up the equipment correctly and are using proper techniques for measuring distance of the spectral lines from the flame.

- Students may have trouble getting enough powdered salt into the flame. An alternative procedure is to sprinkle the powdered salt into the flame from a safe height, or to use a deflagrating spoon to hold a small amount of powder in the flame.

Data and Observations

Table 1 (Sample data are given.)			
Color	Distance D	Distance S	Lines/cm n
Sodium yellow	33 cm	100 cm	5267

Analysis and Conclusions

1. $H = \sqrt{D^2 + S^2} = 105.3$ cm

2. $\sin \theta = \dfrac{33 \text{ cm}}{105.3 \text{ cm}} = 0.313$

3. $d = \dfrac{1}{5276} = 1.9 \times 10^{-4}$ cm

4. Answers will vary. Wavelength values depend on the observed location of the spectral lines.

$\lambda = d \times \sin \theta = 1.9 \times 10^{-4}$ cm $\times 0.313$
$= 5.947 \times 10^{-5}$ cm
$= 595$ nm

Extension and Application

1. The distance D will be less than that for sodium because the wavelength of blue light is less than that of yellow light. To satisfy the Bragg equation at a shorter wavelength, the angle must be smaller, so the distance will be less.

2. Some of the heat energy is absorbed by the electrons in the atoms of the vaporized element. The electrons later release this energy as light. Some heat energy that would be wasted in an ordinary lamp is converted to usable light energy.

29-1

Semiconductor Properties

Introduction

The diode is an active device, since it can control the direction of current flow. A passive device, such as a resistor, only modifies the amount of current flowing in a circuit. Rectifier diodes convert AC current into DC current. Diodes can also control the direction of current flow in circuits, a primitive form of a logic gate. Zener diodes are used in the reverse-bias direction as voltage-control devices. The voltage rating of a semiconductor diode is determined by the materials used in the diode. The voltage rating of a rectifier diode measures how much voltage the diode can sustain in the reverse-bias direction before the diode starts to conduct current in that direction.

The amount of current that can safely pass through a diode is generally a function of the physical size of the diode. In applications where a large current flows through a diode, the heat generated in the diode must be removed.

Process Skills

measuring in SI, making and using graphs, analyzing, observing and inferring, using numbers, communicating, interpreting data

Preparation Hints

- Any rectifier diode can be substituted for the 1N4004 diode as long as the diode working voltage is greater than the voltage you are using. A zener diode should not be used, as it will conduct current in both directions.

Materials

1N4004 diode (rectifier) or equivalent
battery, 1.5–3-V, or DC power supply, 0–5 V
10-Ω resistor
1000-Ω resistor
1000-Ω potentiometer (variable resistor)
power supply, 5 VAC, or AC adapter, such as that used for a tape recorder
connecting wire
knife switch
CBL unit
link cable
voltage and current probe with dual-channel amplifier box
2 CBL-DIN adapters
graphing calculator with PHYSICS CBL software program

- The 10-Ω resistor limits current, to prevent damage to the ammeter in case students apply the full battery or power-supply voltage to the meter. If the power-supply voltage is over 3 V, the resistance can be increased to 33–47 Ω.

- You can use any of the power supplies with an AC power output. The small transformer adapters with an AC output can be substituted. Use only a low-voltage AC power supply, not the 120-VAC line current.

- For Extension and Application, a center-tapped transformer is shown, but most AC adapters and many laboratory power supplies do not provide this feature. If you cannot get such a transformer, simulate a center tap with two equal resistors in series across the secondary. The resistors should have a value between 100 and 200 Ω with a rating of 0.5 W or greater. The junction between the resistors is the simulated center tap. Do not use this setup with a transformer or an adapter with an output voltage greater than 12 V.

Data and Observations

Table 1 (Sample data are given.)									
Voltage (V)	0.00	0.10	0.20	0.30	0.40	0.50	0.60	0.70	0.80
Forward-biased diode, current (mA)	0	0	0	0	0	0	2	17.5	180
Reverse-biased diode, current (mA)	0	0	0	0	0	0	0	0	0

Procedure Hints

- *Caution:* When using the CBL dual-channel amplifier to observe the maximum voltage rating, it is best not to exceed 5 volts.

- Alternative traditional equipment can be substituted for the CBL unit. An ammeter, 0–100 mA or 0–500 mA for the current probe and a voltmeter, 0–5 VDC for the voltage probe can be used instead. An oscilloscope can be hooked up for steps 7 and 8 in the procedure to observe the AC power source and to observe the voltage across the 1000-Ω resistor.

Table 2

Diode with AC power

AC wave shape

AC wave shape through a diode

Analysis and Conclusions

1. Current versus Voltage for Forward and Reverse Biasing of a Diode

Reverse-biased   Forward-biased

Voltage (V)

2. With a resistor in the circuit, a linear relationship between current and voltage exists for both negative and positive values of voltage. With a diode in the circuit, no current flows when it is reverse-biased (for negative voltages), but the current increases exponentially when the diode is forward-biased and the voltage increases.

3. The diode conducts current in only one direction.

4. The diode is silicon. It begins to conduct current with a forward bias at about 0.6 V.

5. An AC source has a sinusoidal wave shape. The wave shape of a current across the resistor that passes through the diode is only the positive portion of the alternating current. The negative portion of the alternating current is blocked.

6. The shape is the opposite half (negative portion) of the AC wave shape that was drawn in Table 2 (AC wave shape through a diode).

Extension and Application

1. The figure shows full-wave rectification. Each diode alternately conducts current for half the time.

2. The wave shape observed should be the same as the one predicted in item 1.

29-2

Integrated-Circuit Logic Devices

Introduction

Integrated-circuit logic devices perform functions from the simple gate functions that students examine here to the very complex devices called microcomputers. The only real difference between the simple and complex devices is the number of gates they have. All IC logic devices are constructed from the simple gate functions. The particular function performed by an IC depends on how its logic elements interconnect.

Misconception: Defining a logical 1 as voltage present is called positive logic. You could just as well define a logical 0 as voltage present; this is called negative logic.

Process Skills

comparing and contrasting; observing and inferring; using variables, constants, and controls; recognizing cause and effect

Preparation Hints

- A kit containing ICs, resistors, LEDs, breadboard, and wires is available from Science Kit.

- Batteries should be used instead of power supplies, which often create problems with ICs.

Materials

breadboard
9-V battery
9-V battery clip
connecting wire, 22 gauge
light-emitting diode (LED)
1000-Ω resistor
CMOS integrated circuits: 4081, 4071, 4001, 4011

Procedure Hints

- Have students check that their probes are working. Students must apply power to the IC or the circuit will not work.

- To simplify the test-probe arrangement, solder the 1000-Ω resistor to one lead of the LED. Then attach two long, sturdy wires to the remaining LED lead and the resistor. Put electric tape or heat-shrink tubing around the wires and the resistor. Label the negative lead. A simple test probe can be constructed by using a two-lead, bicolor LED. A red light indicates current is flowing in one direction, while a green light indicates current is flowing in the other direction. Regardless of the direction of the leads' placement, the LED will light.

- Requiring students to demonstrate their circuits before they continue ensures that each group does its own work. Ask different members to demonstrate different combinations.

Data and Observations

Table 1

AND Gate Truth Table

Inputs		Output
A	B	Z
0	0	0
0	1	0
1	0	0
1	1	1

Table 2

OR Gate Truth Table

Inputs		Output
A	B	Z
0	0	0
0	1	1
1	0	1
1	1	1

Table 3

NOR Gate Truth Table

Inputs		Output
A	B	Z
0	0	1
0	1	0
1	0	0
1	1	0

Table 4

NAND Gate Truth Table

Inputs		Output
A	B	Z
0	0	1
0	1	1
1	0	1
1	1	0

Table 5

NAND Gate Combination

Inputs		Output
A	B	Z
0	0	0
0	1	1
1	0	1
1	1	1

Analysis and Conclusions

1. The combination produces a logical OR.

2. The NOR function is a negative OR (not OR).

3. The NAND function is a negative AND (not AND).

4. The result is an inverter. A logical 1 input produces a 0 output and vice versa.

5. The result is an inverter.

Extension and Application

1. $A = (I \cdot G) \cdot [(L \cdot B_L) + (R \cdot B_R)]$

2.

XOR Gate Truth Table

Inputs		Output
A	B	Z
0	0	0
0	1	1
1	0	1
1	1	0

30-1

Radiation Properties

Introduction

The measured intensity of a radioactive source varies in inverse proportion to the distance between the source and the measuring device. This principle applies to other types of radiation, including light, sound, and radio waves. This principle can be easily visualized if one thinks of a fixed amount of radiation being spread over the surface of an ever-expanding sphere. In this experiment, students will measure the intensity of a radioactive source at increasing distances.

Process Skills

measuring in SI; using variables, constants, and controls; using numbers; making and using graphs; interpreting data

Preparation Hints

- An ionization-type smoke detector, which often contains americium, could be used as a radioactive source. Mantles for gasoline camping lanterns may contain radioactive thorium, which could also be used as a radioactive source.

Materials

Geiger counter or scaler
beta source
tweezers or tongs
gloves, heat-protective
stopwatch or watch with second hand
meterstick
masking tape
graph paper

Procedure Hints

- To facilitate measurements, students can take readings for a shorter time, such as 15 s. The value for that period could be multiplied appropriately to determine the activity in counts per minute.

- Remind students that the decay process is spontaneous and probably will vary during their readings. A smooth curve that best fits the data will take this into account.

Data and Observations

Background radiation = 36 c/min

Table 1 (Sample data are given.)				
Distance (cm)	Measured activity (c/min)	Adjusted activity (c/min)	Distance^2 (cm^2)	1/Distance^2 (cm^{-2})
2	2616	2580	4	0.25
3	1416	1380	9	0.11
4	732	696	16	0.063
5	564	528	25	0.04
6	426	390	36	0.028
7	312	276	49	0.02
8	276	240	64	0.016
9	240	204	81	0.012
10	168	132	100	0.01
11	150	114	121	0.008
12	114	78	144	0.007
13	126	90	169	0.006
14	72	36	196	0.005
15	72	36	225	0.004
16	42	6	256	0.004
17	66	30	289	0.003
18	66	30	324	0.003
19	50	14	361	0.003
20	36	0	400	0.003

Analysis and Conclusions

1. See Table 1.

2.

Activity vs Distance or a Beta Source

This graph was produced with the Lab Partner graphing program.

3.

Activity vs 1/Distance Squared for a Beta Source

4. Background radiation originates from cosmic radiation and from naturally decaying materials in rocks or soil.

5. The graphs show that the intensity of radiation varies inversely with the square of the distance from the source. The graph of adjusted activity versus distance decreases exponentially as the distance increases. The graph of adjusted activity versus the reciprocal of distance squared is a straight line, thus confirming the relationship.

6. The curves have the same shape whether or not the background radiation is subtracted. It is important to subtract the background radiation when the source is weak compared to the background level, or the graph of activity versus distance will be asymptotic.

Extension and Application

1. The results for gamma radiation are similar to those for beta radiation. Sample data:

Background radiation = 26 c/min

Gamma radiation			
Distance (cm)	Measured activity (c/min)	Distance (cm)	Measured activity (c/min)
2	612	12	42
3	360	13	50
4	228	14	42
5	186	15	42
6	120	16	66
7	114	17	24
8	120	18	
9	72	19	
10	78	20	
11	30		

2. Since the radioactive activity decreases at the rate of the reciprocal of the distance squared, the measured activity should be one-ninth of that measured at 5 cm, or about 1100 c/min.

Radioactive Shielding

Introduction

Students investigate the ability of alpha, beta, and gamma radiation to penetrate materials in this experiment. Because alpha particles are physically large and have a positive charge, they interact quickly with the atoms in materials and do not penetrate very well. Beta particles, which are electrons, pass through materials fairly well but attenuate as the thickness increases. The attenua-

tion of beta particles is the result of random collisions with the shell electrons in the atoms of the material. Gamma rays, a form of electromagnetic radiation, pass through many materials with little absorption. In general, only very dense materials having atoms with large nuclei are effective gamma-ray absorbers.

Process Skills

measuring in SI; comparing and contrasting, communicating, observing and inferring

Preparation Hints

- Before starting, check that the alpha source activity is measurable. If the activity is too weak, have students skip this part.

Materials

Geiger counter or scaler
stand for Geiger-Mueller tube
alpha, beta, and gamma sources
5-cm squares of thin cardboard, aluminum, and lead (10 of each)
tweezers, tongs, or gloves
stopwatch or watch with second hand

Procedure Hints

- Organize small groups of three or four. Each student can then be only manipulating layers of shielding material, timing, resetting the Geiger counter, or recording data, and the group can move quickly through the experiment.

Data and Observations

Background radiation = 7 c/10 s

Table 1 (Sample data are given.)								
Number of sheets	Cardboard			Aluminum			Lead	
	alpha	beta	gamma	alpha	beta	gamma	beta	gamma
0	280	670	190	255	484	184	521	173
1	15	661	195	12	450	175	11	160
2	9	656	200	5	409	162	4	147
4		585	173		152	135		114
6		560	154		20	142		120
8		452	162		2	160		96
10		371	169			107		63

Measured Activity (c/10 s)

Analysis and Conclusions

1. Alpha particles are easiest to absorb; gamma rays are hardest to absorb.

2. Cardboard is not an effective absorber of beta or gamma radiation. Beta radiation is reduced to one-half by about three sheets of aluminum. Gamma radiation is reduced to one-half by about eight sheets of lead.

3. Alpha particles are effectively blocked by a pair of cardboard or aluminum sheets, while beta and gamma rays pass through the same amount of shielding. A pair of lead sheets blocks beta rays, while gamma rays pass through the same amount of shielding.

4. The gamma radiation was not eliminated but only reduced as it passed through lead.

Extension and Application

1. Since the intensity of a photon beam decreases exponentially as it penetrates a

particular type of absorber, theoretically, an absorber of infinite thickness would be required to reduce the beam intensity to zero.

2. Beta radiation would be most suitable to measure the thickness of the paper. Alpha radiation would be absorbed by the paper, while gamma radiation would hardly be absorbed at all. The intensity of the beta radiation passing through the paper will be inversely proportional to the thickness of the paper, assuming it has a constant density.

Notes

Notes

Notes

Notes

Notes